Studies in Computational Intelligence

Volume 1123

The series "Studies in Computational Intelligence" (SCI) publishes new developments and advances in the various areas of computational intelligence—quickly and with a high quality. The intent is to cover the theory, applications, and design methods of computational intelligence, as embedded in the fields of engineering, computer science, physics and life sciences, as well as the methodologies behind them. The series contains monographs, lecture notes and edited volumes in computational intelligence spanning the areas of neural networks, connectionist systems, genetic algorithms, evolutionary computation, artificial intelligence, cellular automata, self-organizing systems, soft computing, fuzzy systems, and hybrid intelligent systems. Of particular value to both the contributors and the readership are the short publication timeframe and the world-wide distribution, which enable both wide and rapid dissemination of research output.

Indexed by SCOPUS, DBLP, WTI Frankfurt eG, zbMATH, SCImago.

All books published in the series are submitted for consideration in Web of Science.

Animesh Mukherjee · Juhi Kulshrestha ·
Abhijnan Chakraborty · Srijan Kumar
Editors

Ethics in Artificial Intelligence: Bias, Fairness and Beyond

Springer

Editors
Animesh Mukherjee
Department of Computer Science
and Engineering
IIT Kharagpur
West Bengal, India

Abhijnan Chakraborty
Department of Computer Science
and Engineering
IIT Delhi
New Delhi, Delhi, India

Juhi Kulshrestha
Department of Computer Science
Aalto University
Espoo, Finland

Srijan Kumar
School of Computational Science
and Engineering, College of Computing
Georgia Institute of Technology
Atlanta, GA, USA

ISSN 1860-949X ISSN 1860-9503 (electronic)
Studies in Computational Intelligence
ISBN 978-981-99-7183-1 ISBN 978-981-99-7184-8 (eBook)
https://doi.org/10.1007/978-981-99-7184-8

This Springer imprint is published by the registered company Springer Nature Singapore Pte Ltd.
The registered company address is: 152 Beach Road, #21-01/04 Gateway East, Singapore 189721, Singapore

Paper in this product is recyclable.

Foreword

In an era marked by unprecedented advancements in technology, the ethical considerations surrounding Artificial Intelligence have taken center stage. The relentless march of AI, with its promise of transformative power, has left no stone unturned in reshaping our world. It is a phenomenon that holds immense potential, yet simultaneously triggers an avalanche of concerns.

In the grand symphony of AI, where lines between science fiction and reality blur, we find ourselves at the intersection of innovation and moral responsibility. The book you are about to embark upon, "*Ethics in Artificial Intelligence: Bias, Fairness and Beyond*," ventures into the heart of this dilemma, skillfully unraveling the multifaceted aspects of ethics in AI.

But this book is more than just an exploration of the burgeoning ethical landscape in AI. It is a collective endeavor by some of the brightest minds in the field, individuals who have dedicated their knowledge and insights to dissect the profound issues of bias, discrimination, fairness and interpretability within the domain of algorithmic AI systems.

As we delve into the chapters, we confront pressing questions: How do we combat the insidious specter of biased AI behavior, entrenching discriminatory tendencies in human decision-making? How do we attribute accountability for accidents and errors caused by AI systems? Can AI systems coexist harmoniously with human intelligence to build a more equitable and sustainable world?

Through the contributions of experts from academia and industry, this volume weaves together a tapestry of ideas, solutions and illuminations. From the importance of fairness in software testing to the delicate balance of ethical considerations in computational social science, the chapters offer a panoramic view of the challenges and prospects AI brings to the table.

In this age of remarkable AI advancements, where it's celebrated for its capacity to push the boundaries of human knowledge, but at the same time feared for its potential to subsume our ethical underpinnings, this book is a beacon, a timely reminder of the ethical obligations that accompany the AI revolution.

I would like to express my gratitude to the esteemed editors and authors who have assembled this profound work. Their collective wisdom and dedication have resulted in a volume that is both thought-provoking and inspiring.

The Institution of Engineers (India), under the aegis of the WFEO Committee on Information and Communication (WFEO-CIC), is proud to bring you this book. It serves as a testament to our commitment to the ethical progression of AI. As we usher in an age where AI's influence on society is undeniable, it becomes imperative to equip ourselves with the critical thinking skills necessary to discern the ethical nuances that permeate this technology.

"*Ethics in Artificial Intelligence: Bias, Fairness and Beyond*" is not just a book; it is a guide, an intellectual compass that will help you navigate the complexities of the AI landscape. It's a reminder that the future of AI must be steered with a steady hand, grounded in ethical principles.

As you embark on this enlightening journey through the pages of this book, may you find the insights and perspectives you seek to understand the evolving world of AI and its profound societal implications.

Let the voyage begin.

Gujarat, India

S. S. Rathore
Chair, WFEO-CIC

Preface

This book introduces some of the latest developments in the area of ethical practices in Artificial Intelligence. As AI becomes pervasive in physical and online domains to serve many crucial applications and decision-making systems, such as giving bail, self-driving, providing healthcare recommendations and more, it is of critical importance to understand the potential downsides and the unintended consequences of AI. While pioneers like Ray Kurzweil paint a very promising picture of the future of AI, philosophers like Nick Bostrom caution us about the emergence of "super intelligence". In fact, visionaries like Elon Musk, Bill Gates and Stephen Hawking have repeatedly urged administrations to factor in ethical and engineering standards in the development and deployment of AI systems that are going to somehow "monitor" the future of humanity. Some of the glaring questions of current times are—How can we tackle the problem of biased learning in AI behavior; for instance, reinforcing racial and gender informed discriminations in human decision-making? How do we allocate responsibility for accidents/errors caused by AI systems such as autonomous vehicles or medical diagnostics? Can we have AI systems that complement rather than displace human intelligence in order to support a more sustainable and just world?

This book seeks to answer some of these important yet hard questions posed above. Leading experts from academia and industry have contributed together to bring forth some of the most intriguing issues around bias, discrimination, fairness and interpretability in the design of algorithmic AI systems. Saha and his co-authors present an account of the importance of fairness considerations in software testing, debugging and repairing. Next, Gupta, Jain and their co-authors demonstrate how to ensure individual and group fairness in the design of clustering algorithms. Gupta et al. subsequently discuss fairness issues over time. Later, Makhortykh discusses the impact of memory ethics in the retrieval of genocide-related information. While Patro discusses fairness in multi-party systems, De and her co-authors illustrate the need for ethical thinking in designing algorithms for computational social science. This is followed by a very interesting chapter on the design of fair allocation by Biswas and her co-authors. The volume ends with a chapter by Sikdar et al. discussing the significance of the design of interpretable and explainable machine learning models.

While AI has become one of the most "glamorous" words in the technology world and academicians, as well as practitioners, are in a race to be at the forefront of this AI movement, less attention has been paid to the repercussions of this movement. This volume is geared toward increasing awareness of the immediate moral, ethical and legal repercussions of the presence and possibly dominance of AI in our society. As AI starts to influence society, and in turn, is influenced by inputs from society, this volume seeks to impart to the reader the critical thinking skills necessary to understand the pros and cons that applications of AI can have on the broader audience, including the marginalized communities that may be impacted by the AI system. Further, this volume will discuss potential ways to improve the ethical implications of the AI models, such as making them interpretable. The book volume is meant to investigate the level of truth in the possibility of AI emulating consciousness, cognition, conation and emotion in an "artificial being" in the future, and if so, then what would be the implications of that possible reality. The main aim is to prepare the audience to understand the critical capacity of the rapidly evolving AI technology and its societal implications.

We acknowledge the painstaking contributions of our esteemed referees without whose support we would not have been able to deliver the book at its finest quality. We thank Soumya Sarkar, Gaurav Verma, Indira Sen, Abhisek Dash, Anjali Gupta and Takayuki Hiraoka for their unconditional help. We also thank The Institution of Engineers India for facilitating this volume.

West Bengal, India — Animesh Mukherjee
Espoo, Finland — Juhi Kulshrestha
New Delhi, India — Abhijnan Chakraborty
Atlanta, USA — Srijan Kumar

Contents

About the Editors

Animesh Mukherjee is an Associate Professor and the A. K. Singh Chair at the Department of Computer Science and Engineering, IIT Kharagpur, West Bengal, India. His main research interests center around investigating hate and abusive content on social media platforms, fairness, bias in information retrieval systems, media bias, and quality monitoring of Wikipedia articles. He publishes and is on the committee of most of the top AI conferences, including The Web Conference, NeurIPS, AAAI, IJCAI, ACL, EMNLP, NAACL, Coling, CSCW, ICWSM, etc.

Juhi Kulshrestha is an Assistant Professor at the Department of Computer Science at Aalto University, Finland. She obtained a doctoral degree from the Max Planck Institute for Software Systems, Germany. She also pursued research at the University of Konstanz and Leibniz Institutes for Social Sciences and Media Research. Her research, at the intersection of computer science and social science, broadly focuses on leveraging digital behavioral data to quantify and characterize how people consume news and information on the web and its effect on society. She regularly publishes in and serves on the committees of top-tier venues such as TheWebConf, AAAI ICWSM, ACM CSCW, and ACM FAccT. She is a recipient of several internationally competitive research grants such as Meta's Foundational Integrity Research Grant, Social Science One's Social Media and Democracy Research Grant, Google European Doctoral Fellowship for Social Computing, and Google Anita Borg Scholarship.

Abhijnan Chakraborty is an Assistant Professor at the Department of Computer Science and Engineering, Indian Institute of Technology (IIT) Delhi. He is also associated with the School of Artificial Intelligence and the School of Information Technology at IIT Delhi. His research interests fall under the broad theme of Computing and Society, covering the research areas of Social Computing, Information Retrieval and Fairness in Machine Learning. In the past, he has worked at the Max Planck Institute for Software Systems, Germany and Microsoft Research India. During Ph.D., he was awarded the Google India Ph.D. Fellowship and the Prime Minister's Fellowship for Doctoral Research. He regularly publishes in top-tier computer science conferences including WWW, KDD, AAAI, AAMAS, CSCW and ICWSM. He has won INAE Young Engineer 2022 award, the best paper award at ASONAM'16 and best poster award at ECIR'19. He is one of the recipients of an internationally competitive research grant from the Data Transparency Lab to advance his research on fairness and transparency in algorithmic systems.

Srijan Kumar is an Assistant Professor at the School of Computational Science and Engineering, College of Computing at the Georgia Institute of Technology. He completed his postdoctoral training at Stanford University, received a Ph.D. and M.S. in Computer Science from the University of Maryland, College Park, and B.Tech. from the Indian Institute of Technology, Kharagpur. He develops Data Mining methods to detect and mitigate the pressing threats posed by malicious actors (e.g., evaders, sockpuppets, etc.) and harmful content (e.g., misinformation, hate speech etc.) to web users and platforms. He has been selected as a Kavli Fellow by the National Academy of Sciences, named as Forbes 30 under 30 honoree in Science, ACM SIGKDD Doctoral Dissertation Award runner-up 2018, and best paper honorable mention award from the ACM Web Conference. His research has been covered in the popular press, including CNN, The Wall Street Journal, Wired, and New York Magazine.

Testing, Debugging, and Repairing Individual Discrimination in Machine Learning Models

Diptikalyan Saha, Aniya Agarwal, Sandeep Hans, and Swagatam Haldar

Abstract Trustworthy AI is crucial for the broad adoption of AI systems. An important question is, therefore, how to ensure this trustworthiness. The absence of algorithmic bias is a crucial attribute for an AI system to be considered trustworthy. In this book chapter, we address various problems related to the detection and mitigation of algorithmic bias in machine learning models, specifically individual discrimination. A model shows individual discrimination if two instances, predominantly differing in protected attributes like race, gender, or age, produce different decision outcomes. In a black-box setting, detecting individual discrimination requires extensive testing. We present a methodology that enables the automatic generation of test inputs with a high likelihood of identifying individual discrimination. Our approach unites the power of two widely recognized techniques, symbolic execution and local explainability, to generate test cases effectively. We further address the problem of localizing individual discrimination failures required for effective model debugging. We introduce a notion called Region of Individual Discrimination or RID, which is an interpretable region in the feature space, described by simple predicates on features, aiming to contain all the discriminatory instances. This region essentially captures the positive correlation between discriminatory instances and features. Finally, we describe a model repair algorithm that aims to repair individual discrimination in the model by effectively creating retraining data based on RID. We empirically show that our approaches are effective for assuring individual fairness of machine learning models.

1 Introduction

The last decade has seen a great increase in the adoption of AI models for decision-making systems. We have seen applications of AI in some critical decision-making including loan prediction [27], churn prediction [20], criminal justice [12], graduate

D. Saha (✉) · A. Agarwal · S. Hans · S. Haldar
IBM Research, Bangalore, India
e-mail: diptsaha@in.ibm.com

A. Mukherjee et al. (eds.), *Ethics in Artificial Intelligence: Bias, Fairness and Beyond*,
Studies in Computational Intelligence 1123,
https://doi.org/10.1007/978-981-99-7184-8_1

admission [4], and healthcare [15]. Due to the uninterpretability of complex AI models, trustworthiness is always under scrutiny. Industrial usage of AI models needs a strong development and maintenance life cycle which can ensure trustworthiness in AI models. Similar to the Software Development life cycle, automation in the AI life cycle is of utmost importance. Various manual tasks for developers in AI/ML developer life cycle like data preprocessing and hyperparameters optimization are already getting automated by AutoML [1] tools like TPOT [24]. This work is a step toward improving developer productivity in the AI life cycle, similar to various debugging and fault localization tools developed in non-AI software engineering.

One of the critical dimensions where AI models seem to have failed in the past is fairness. Several instances [25] of such critical failure raise an important question—how can we ensure that the model is fair before it is put to production usage? There are three important questions we try to address—how do we generate effective test cases, how do we localize the faults in the feature space, and finally how do we repair the models?

There are two types of fairness that have mainly been of concern to society—group discrimination and individual discrimination. Group discrimination concerns with disparate impact [16] and disparate treatment [25] of two groups (e.g., male/female, black/white) defined by the protected attributes (gender, race, etc). While, individual discrimination [14, 17] deals with discrimination at the individual input level where two inputs differ mainly in their protected attribute values, but still receive different predictions by the model. As group discrimination can be addressed without parity at the individual level and handling individual discrimination does not ensure group fairness as groups may not have similar individuals, these two problems remain important in their own merit [5]. In this book chapter, we scope our investigation to testing, localizing, and repairing of *individual discrimination* for *black-box classification models built on tabular data*.

The goal of testing individual discrimination is to generate *diversified* and *effective* test samples. Diversification ensures that the test samples are distributed in feature space and effectiveness assigns a high probability of failures to those test cases. We aim to systematically search feature space for better coverage without much redundancy. In software engineering literature, symbolic execution-based techniques are available [7, 19, 32] for the automated generation of test inputs, systematically exploring various execution paths within a program. These methods prevent the generation of redundant inputs that traverse the same program path. They have been successfully employed to generate inputs for interpretable procedural programs. In this chapter, we use the local interpretable model like a decision tree arising out of a local explainer such as LIME [30] to perform symbolic execution. We also use the local explainer to perturb only those non-protected attributes which have more influence on the prediction, thereby forcing the generation of more discriminatory test cases.

Effective test case generation techniques such as the one above do not answer the following question—*What features are positively correlated with the individual discrimination failures?* We answer this question in this chapter by *identifying interpretable constraints on input space that are positively correlated with discriminatory behavior.* Such constraints on the input space localize the discriminatory behavior to an interpretable region, called the Region of Individual Discrimination, in short RID. This information is significantly valuable in the following scenarios:

- *Check and Debug:* A highly-dense discriminatory region in the training data would signify that the problem is with the model and debugging effort should be concentrated on hyperparameter optimization or choosing a different model. If the discriminatory region is less dense, then training data augmentation or gathering needs to be done.
- *Generate and Test:* The discriminatory regions can be used to generate more discriminatory inputs with high precision/effectiveness that can be used for further testing.
- *Guardrails:* The discriminatory regions can be used to know a-priori if the input is going to be discriminatory (with high probability). This can help in taking appropriate action (e.g., creating firewall rules to divert the traffic) on the input regions in deployment.

Once model testing finds discriminatory samples, it is important to repair the model. Most fairness mitigation strategies can be classified into preprocessing (changing training data), in-processing (changing model construction), and post processing (changing model output). In this case, we present a mitigation strategy that changes the training data for retraining the model. The solution objective is to generate repairing data that can reduce the discriminatory behavior of the model without affecting the generalization ability. We present an iterative algorithm to minimize the objective.

The rest of the chapter is organized as follows. Section 2 presents the required background on individual discrimination and decision tree. Sections 3, 4, and 5 present the test generation, fault localization, and repair algorithms, respectively. Further, in Sect. 6, we present the experimental results. Next, we present some related work in Sect. 7, while Sect. 8 describes a brief summary along with the future direction of this work.

2 Background and Notation

2.1 Fairness

In this section, we present some related concepts that we use in our framework.

Fairness. A *fair* classifier tries not to discriminate among individuals or groups defined by the *protected attribute* (like race, gender, caste, and religion) [17]. The protected attribute with a value indicating a group that has historically been at a systematic advantage is the privileged group, while the rest are called unprivileged.

In a binary classification scenario, the positive class corresponds to the favorable outcome that individuals wish to achieve, whereas the negative class corresponds to an unfavorable one.

Notation. Say $\mathcal{D} = (X, Y)$ is a data distribution where X is the input domain/space with set of features F (numerical or categorical) and Y is the label space. Tr and Te are two subsets of $\mathcal{D}$ used to train and test a classifier model M, respectively. We use x, x' to denote an *input* from the input space X. We use the term *instance* (i.e., containing both input and label) as an element of the set $\mathcal{D}$. For any instance set D', we use D'_X to denote its inputs and D'_Y to denote its labels. The classifier M is a function $M : X \to Y$ and the final prediction of input $x \in X$ is denoted by $\hat{M}$ where $\hat{M}(x) = 1 \Leftrightarrow M(x) > threshold$ and 0 otherwise. Let $P \subset F$ denote the set of all protected features.

Individual Fairness There exist many definitions/measures for individual fairness [14, 17, 21–23]. Dwork et al. [14] states that any two individuals with similar non-sensitive features should receive similar predicted results. Based on the notion of counterfactual fairness [22], a decision is fair toward an individual if it is the same in (a) the actual world and (b) a counterfactual world where the individual belonged to a different demographic group. Galhotra et al. [17] say two individuals differing in their protected attribute(s) should get the same outcome. Given such a variation in the definition, we designed our algorithms to be agnostic to these definitions, i.e., our algorithm does not exploit any property related to how different definitions determine similar individuals and compare their predictions. We use $I(x, x', M, P)$ to generically denote if two similar inputs x and x' are discriminated by model M with respect to the protected attributes P under some fairness definition.

An input x is *discriminatory*, denoted as $D(x, P, M)$ with respect to protected attributes P and model M, if there exists an input x' such that $I(x, x', M, P)$ holds, i.e.,

$$D(x, P, M) = 1 \Leftrightarrow \exists x' \in X \text{ such that } I(x, x', M, p) \text{ is } True \qquad (1)$$

We assume that each discrimination notion comes with a function to efficiently search an x' for a given x s.t. $I(x, x', M, P)$ holds. As per Dwork's definition, this will amount to searching in a neighborhood of x defined by a distance function and checking if the predictions for x and x' differ. With THEMIS [17], this amounts to only perturbing the protected-attributes values to generate x' and checking if the prediction changes i.e. $\hat{M}(x) \neq \hat{M}(x')$.

Since we will be considering analysis for a fixed set of protected attributes P and for a particular model M, we will omit that context P and model M and denote a discriminatory input x as $D(x) = 1$. Given a set of inputs $X_1 \subset X$, $disc(X_1, P, M)$ denotes the set of discriminatory inputs in X_1, i.e., $disc(X_1, P, M) = \{x \in X_1 | D(x, P, M)\}$.

Note that bias removal is not possible by preprocessing the training data to remove the protected attributes as possible correlations may still remain between protected and non-protected attributes. The Adult income [26] dataset has such a correlation between race and zip code.

2.2 Decision Tree

This paper extensively uses decision tree [6]-based algorithms, and therefore, we briefly describe the algorithm next. We have used a variation of the algorithm (available in *Scikit-Learn* [28]) where the user can specify an extra parameter $min\_samples\_leaf$ along with the training inputs. A decision tree classifier checks if the inputs are pure (i.e., have the same class labels) and terminate after considering them as a leaf with its label as the class label. In case of an impure leaf, it decides to further split the inputs unless the sample count there is less than $min\_samples\_leaf$. The splitting is performed by iterating through all features and considering different split values and then choosing the split value which results in maximum information gain in the resultant two splits. The feature with the most effective split condition is chosen as the predicate at that decision tree node. The process continues recursively for the examples in each partition. Scikit-Learn produces a pure decision tree when $min\_samples\_leaf$ is not present in the argument.

3 Testing

In this section, we discuss our overall solution divided into three subsections. The first subsection discusses the aims of our test case generation under different scenarios. We have developed our algorithms by combining two different search techniques—global and local search. We discuss them in the following two subsections.

3.1 Problem Setup

We aim to achieve the following two optimization criteria through the devised test case generation technique.

Effective Test Case Generation: The objective is to generate test cases that maximize the effectiveness ratio, i.e., $|Succ|/|Gen|$, where a model, a set of domain constraints, and a protected attribute set are given. Here, Gen represents the set of non-protected attribute value combinations generated by the algorithm, and $Succ$ is a subset of Gen consisting of instances that result in discrimination, meaning each instance in $Succ$ produces different decisions for varying combinations of protected attribute values.

Here are a few key points regarding this criterion.

- *Test case*: Instead of considering each test case as a collection of values for all attributes, only the non-protected attributes are taken into account. By adopting this approach, we can ensure that the same combination of non-protected attribute values does not contribute to the count of multiple discriminatory test cases.
- *Domain constraints*: It is assumed that the application of domain constraints C will effectively eliminate unrealistic or invalid test cases.
- *Ordering generation and checking*: The above-discussed optimization criteria do not provide specific guidance on whether test cases should be generated all at once (offline) or if discrimination checking and generation can be performed concurrently (online). As a result, the test case generation process can be designed to incorporate discrimination checking in real time, allowing for a more dynamic approach where the generation of test cases depends on the ongoing assessment of discrimination.

In the field of software testing, numerous predefined white-box coverage criteria are available. These criteria have also been extended to encompass machine learning in recent studies [34]. Now, we introduce the path coverage criteria, which have been defined in a manner that makes them applicable to diverse types of models.

Coverage criteria: It should be noted that defining path coverage criteria for black-box models is a complex task. The process of defining *paths* for different types of models relies on their specific operational characteristics. For example, in a neural network, it is possible to define a path based on the activation of neurons, similar to how branch coverage is defined in other contexts [34]. Similarly, in a decision tree classifier, decision paths can be identified and considered for coverage analysis.

Our coverage criteria are defined as follows: Given a set of test cases and a classification model, the coverage is determined by the total number of decision regions executed by the test cases.

In this chapter, a decision tree classifier is employed to approximate the behavior of the model. A highly accurate decision tree model is generated to effectively approximate the decision regions of the given model.

The objective of our test case generation technique is to optimize both path coverage and individual discrimination detection. However, it is important to acknowledge that there are practical limitations to the automatic test case generation process. In our approach, we specifically consider two potential limits: (1) the maximum number of generated test cases, and (2) the time taken to generate them. These constraints ensure that the generation process is performed within reasonable bounds while still striving to achieve the desired objectives.

In the following subsections, we introduce our algorithm that aims to maximize path coverage and successfully detect discrimination. We also describe the approach we use to effectively combine these two objectives.

3.2 Maximizing Path Coverage

To maximize path coverage, we utilize the capabilities of the symbolic execution algorithm. This algorithm systematically explores various execution paths within the model to generate test cases. By leveraging symbolic execution, we can effectively traverse different paths and achieve a comprehensive coverage of the model's behavior. In this subsection, we will discuss a generic symbolic execution algorithm and then modify the algorithm for testing individual discrimination in ML models.

Automated test generation by Dynamic symbolic execution (DSE) [7, 19, 32] proceeds as follows: the engine first instruments the program and then runs the program the instrumented code collects the constraints related to each path. Once an execution path is obtained, the engine will systematically negate the predicates occurring in the path constraint. This essentially generates a new set of path constraints whose solution generates new inputs to further explore a new path.

The method explores each path in a systematic way, and therefore, does not produce any redundant inputs that will try to explore the already executed path. Path coverage criterion aims to execute every possible path at least once.

Next, we formally present a generalized version of the above algorithm in Algorithm 1 which can be used as a framework to generate test cases for AI models. The algorithm starts with one or more seed inputs (Line 3). Then it considers each input and finds the path after execution (Line 9). It then toggles various predicates in the path (Lines 11–19) to generate various constraints as illustrated in Fig. 1. Note that the other variables, not present in the constraint, can take any value from its possible domain. It then checks if a path constraint is already visited or not to ensure that an already traversed path is not traversed again. After creating input by solving the path constraints using a constraint solver, it then uses a ranking function to select which input to process next.

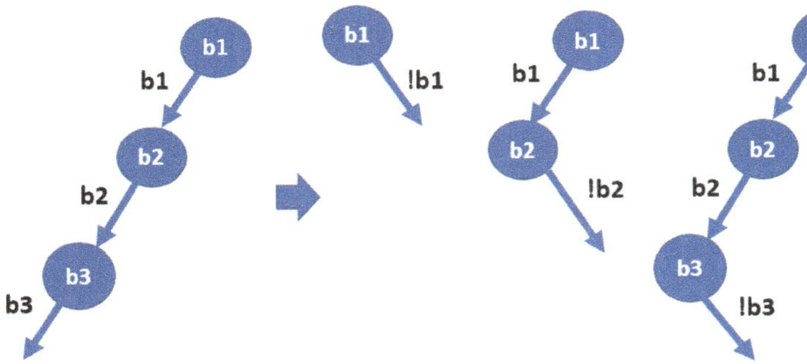

Fig. 1 Generation of path constraints. The three path constraints are created by respectively toggling the branch conditions in the first path

Algorithm 1: Generalized Symbolic Execution

```
count = 0;
covered_path = empty
inputs = initial_test_inputs()      ▷ Seed inputs
pq = empty;            ▷ Priority Queue
pq.addAll(inputs,0)          ▷ Enqueue with priority
while count<limit && !pq.isEmpty() do
    t = pq.dequeue()
    error_condition_check(t)
    p = get_path(t)
    pred_prefix = true
    foreach predicate cons in sorted order from top of path p do
        pc = pred_prefix ∧ toggle (cons)
        if !covered_path.includes(pc) then
            covered_path.add(pc)
            path_input= solve_constraint(pc)
            rank = ranking_algorithm(pc)
            pq.add(path_input,rank)
        end
        pred_prefix = pred_prefix ∧ cons
    end
    count++
end
```

In this subsection, we describe the various functions that are kept undefined in Algorithm 1. The final algorithm is presented in Algorithm 2. Maximizing the path coverage is done in the global search module as mentioned in our final algorithm presented in Algorithm 2.

Path Exploration. We will begin by looking at the `get_path` method. Our main methodology relies on the set of inputs and outputs generated by the local explainer, LIME. However, unlike the original implementation of LIME which learns a linear regressor, we further fit a decision tree on the set of inputs and outputs (refer to Line 40 in Algorithm 2).

The Local Interpretable Model-agnostic Explanation (LIME) [30] is an explanation technique that gives a faithful and interpretable explanation for a prediction from a model, regardless of what kind of classifier or regressor it is. LIME accomplishes this by approximating the true model as an inherently interpretable model locally around the prediction. Inherently interpretable models, such as sparse linear models, shallow decision trees, or falling rule lists, can provide users with easily understandable explanations through visual or textual artifacts. LIME generates data points near a given input data instance by perturbing the input and observing the corresponding output from the original model. By leveraging this input–output information, LIME learns an interpretable model (sparse linear regressor) that maximizes local fidelity and interpretability.

Algorithm 2: Individual Discrimination Testing

```
function generate_test_suite(seed_input, model, cluster_size, limit, Rank1,
  Rank2, Rank3):
    count = 0;
    covered_path = { }
    pq = initial_test_input(seed_input, cluster_size, limit, Rank1)
    while count<limit && !pq.isEmpty() do
        t = pq.dequeue()
        found = D(t)
        p = get_path(model, t)
        if found then
            // local search
            foreach predicate cons in sorted order from the top of path p do
                path_constraint = p.constraint
                If cons is of protected attribute then continue
                path_constraint.remove(cons) path_constraint.add(not(cons))
                if !covered_path.contains(path_constraint) then
                    covered_path.add(path_constraint);
                    input = solve_constraint(path_constraint,t)
                    r = average_confidence_ranking(path_constraints)
                    pq.add(d, Rank2+r)
                end
            end
        end
        // global search
        pred_prefix = true
        foreach predicate cons in sorted order from the top of path p do
            If cons is a protected attribute then continue
            If cons.confidence < T1 then break
            path_constraint = pred_prefix ∧ not (cons)
            if !covered_path.contains(path_constraint) then
                covered_path.add(path_constraint);
                input= solve_constraint(path_constraint)
                r = average_confidence_ranking(path_constraints)
                pq.add(input,Rank3+r)
            end
            pred_prefix = pred_prefix ∧ cons
        end
        count++
    end
    return
function get_path(model, input):
    input_output = LIME-based_local_explanation(model,input)
    decision_tree = train_decision_tree(input_output)
    return decision_tree.path(input);
```

By utilizing the `get_path` method described above, the algorithm operates similarly to the symbolic execution algorithm. It has the ability to negate conditions found within the decision tree path, resulting in the creation of a fresh set of constraints.

These constraints can then be solved using a constraint solver, generating additional inputs that can explore alternative paths.

The application of symbolic execution and local model approximation to a path encounters three primary challenges. The first two challenges stem from the inherent approximation involved in the local model, while the third challenge arises from symbolic execution.

- *Approximation*:
  The decision tree path provides an approximation of the actual program execution path using predicates on features. This approximation can lead to the generation of duplicate program paths.
- *Confidence*:
  Each predicate within the decision tree path is associated with a confidence score, unlike a condition in a program path. Utilizing this confidence score poses a challenge in effectively exploring the paths.
- Symbolic execution in software testing faces the issue of path explosion [18], particularly when using a depth-first-search approach. It can continuously explore paths deep within the program tree while neglecting other parts of the program. Researchers have proposed various techniques to tackle this problem. Demand-driven or directed techniques generate test cases targeted at specific locations in the program, while compositional techniques analyze individual functional modules separately before combining them to generate longer paths throughout the entire program. These techniques leverage the structure of the program being tested.

Path explosion problem.
In order to address the issue of path explosion and avoid focusing solely on a limited portion of the overall space, we utilize the inherent distribution within the available data (training or testing data). Each data instance can serve as a suitable starting point for the symbolic search.

However, when a search limit is imposed and executing all data instances is not feasible, the order of data instances becomes crucial. To enhance diversity in the search process, we employ data clustering techniques and process seed inputs from each cluster in a round-robin fashion.

Local Model Approximation.
To acquire an interpretable local model, we employ a pre-existing local explainer called LIME. The specific perturbation technique used in LIME falls outside the scope of our algorithm, making it challenging to reduce errors caused by local approximation. It's important to note that this approximation may result in the generation of test inputs that fail to reveal any new paths in the model. However, our algorithm effectively resolves this concern by introducing data instances as seed inputs. Moreover, our algorithm prioritizes seed inputs over inputs generated by the constraint solver during symbolic execution, ensuring comprehensive coverage of diverse paths.

Confidence-based Ranking.
When conducting automated test case generation, it is common to have a predefined limit. In such cases, it is crucial to generate test inputs that are non-redundant and effectively cover as much of the path as possible within the given limit. To determine the execution order of test cases generated by the constraint solver, we employ a ranking scheme based on the confidence levels of predicates in the decision tree. This scheme assists in selecting which test input to execute next. The confidence of a path is computed by calculating the average confidence of all predicates present within it (as shown in Line 32, Algorithm 2). Consequently, when the algorithm chooses a predicate, denoted as c, to negate, the average confidence of the predicates within the path leading to and including c is considered as its rank.

Confidence Threshold.
A predicate's confidence level within the decision tree path directly impacts the generation of diverse inputs. If the confidence is low, the likelihood of producing varied inputs decreases. In order to avoid generating unnecessary inputs through symbolic evaluation, our algorithm incorporates a threshold on the predicate's confidence for selection (as shown in Line 27, Algorithm 2). The specific value for this threshold is determined through experimentation.

3.3 Maximizing Effectiveness of Discrimination Detection

Now we explore additional modifications implemented in the generic algorithm to enhance the detection of individual discrimination.

Checking Individual Discrimination. We will begin by discussing the process of checking individual discrimination, which is implemented in the method `check_for_error_condition` as outlined in Algorithm 2. This algorithm is designed to verify individual discrimination based on the defined criteria. In this context, a test case is considered individually discriminatory if, while keeping the values of its non-protected attribute set constant, altering the values of its protected attribute set to every possible combination results in different class labels.

Local Search. As mentioned in previous sections, symbolic execution is employed to discover test inputs that maximize path coverage. This symbolic search strategy, referred to as the *global search*, aims to generate diverse test inputs. Within this approach, certain test inputs generated through seed data or symbolic execution may exhibit discriminatory behavior. To enhance the likelihood of generating discriminatory test cases, we utilize the capability to execute test cases and subsequently check for discrimination. Based on the discrimination analysis, we can generate additional test cases accordingly.

Once a discriminatory test case, denoted as t, is discovered, we proceed to generate additional test inputs that have the potential for individual discrimination. The idea is to negate the non-protected attribute constraints within the decision tree of test case t to generate new test inputs. By toggling a single constraint associated with a non-protected attribute and generating an input that satisfies the resulting constraint, the algorithm explores the neighborhood of the discriminatory path p. This approach, referred to as *local search*, focuses on searching the vicinity of discriminatory test cases. It allows for the exploration of nearby paths that may exhibit individual discrimination characteristics.

This approach is effective due to the inherent adversarial robustness property of machine learning models. It has been observed that even a small perturbation in the input can lead to a change in the classifier's decision [35].

Sticky Solutions. The primary objective of both local and global search is to explore a wide range of paths within the model. The local search focuses on investigating paths that are in close proximity to discriminatory paths, which are generated from discriminatory inputs. As a result, the local search typically yields a single solution for the constraint. However, to address potential approximations introduced by the local linear model, we utilize a *sticky solution* approach. A sticky solution refers to a solution obtained from the constraint solver that closely aligns with the solution of the previous constraint, which was associated with the discriminatory input. This stickiness ensures that when one predicate is negated, the remaining predicates tend to retain the same values as in their previous solutions. By leveraging sticky solutions, we can maintain a level of consistency and coherence within the generated test cases.

The algorithm for test case generation to detect individual discrimination is outlined in the method `generate_test_cases` as shown in Algorithm 2. The local search is described in Lines 11–22, while the global search is described in Lines 24–35. There are two main differences between the implementations of the local and global search. Firstly, in the local search, there is no threshold-based constraint selection. In other words, all constraints are considered for toggling, without any threshold criteria. This approach aims to maximize the coverage of different paths, allowing for a more comprehensive exploration of the vicinity of discriminatory paths. Secondly, in the global search, there is no constraint defined for the suffix of the path. This means that only the constraints leading up to a certain point in the path are considered for toggling. In contrast, during the local search, all constraints, except the selected low confidence one to toggle, remain unchanged. This distinction ensures that the local search focuses specifically on exploring the neighborhood of discriminatory test cases, while the global search takes a broader approach. By employing these differences in implementation, the algorithm aims to strike a balance between comprehensive path coverage and targeted exploration of discriminatory paths.

Ordering Local and Global Search. In the Algorithm 2, three reference ranks, namely $Rank1$, $Rank2$, and $Rank3$, are defined to prioritize the different sources of discriminatory inputs. These ranks are assigned in a way that gives the highest priority to the local search, followed by the seed input, and then the global search, based on their effectiveness in uncovering inputs that lead to discrimination. At Line 4, the rank

*Rank*1 is assigned to the inputs discovered during the local search. This indicates that the discriminatory inputs found through the local search are given the highest priority. Similarly, at Line 19, the rank *Rank*2 is assigned to the seed inputs. These are the initial inputs used for test case generation and are considered to have a relatively lower priority than the inputs found through the local search. Finally, at Line 33, the rank *Rank*3 is assigned to the inputs obtained during the global search. These inputs are considered to have the lowest priority in terms of uncovering discrimination. By assigning these reference ranks, the algorithm ensures that the local search, which is expected to yield the most discriminatory inputs, is given the highest priority. The seed inputs and the inputs from the global search are considered secondary and tertiary, respectively, in terms of their potential to uncover discrimination-causing inputs.

4 Debugging

In this section, we address the problem of identifying *RIDs* which localizes the discriminatory behaviors of the classifier. We aim to express these regions in an interpretable fashion as the conjunction of predicates on attributes, e.g., (age > 30 AND salary > 10000) without containing the protected attributes. One important observation which motivates identifying the discriminatory regions is that the discriminatory inputs exist in close proximity to each other to form a region as shown in Fig. 2.

There are a few important challenges—(1) the RIDs should be precise (most inputs in the regions are discriminatory) and (2) high-recall (all discriminatory inputs are covered using the RIDs). The higher precision will increase the efficiency of finding discriminatory inputs. In addition to higher precision, a higher recall will ensure that at runtime, it can be used to predict if the payload input will be discriminatory. (3) The last challenge that we address is the generation of the RIDs using as few inputs as possible.

Approach. One RID can be interpreted as a conjunction of predicates on the attributes. A classification model that is suitable to create such a boundary is a decision tree [6]. Therefore, the naive approach to solve the problem is to determine the discriminatory and non-discriminatory inputs in the available seed data (e.g., training or validation data of the target model) and treat it as training data for creating the decision tree model which can distinguish the discriminatory and non-discriminatory inputs.

Our main contribution is to extend this simple approach in two ways. The first extension is based on an important observation related to individual discrimination. Individual discrimination is a metamorphic property that does not need a human to provide the ground truth. In other words, we can automatically determine if an arbitrary input is discriminatory. Therefore, it is possible to augment the seed data of the decision tree model by querying the target model for synthetically generated

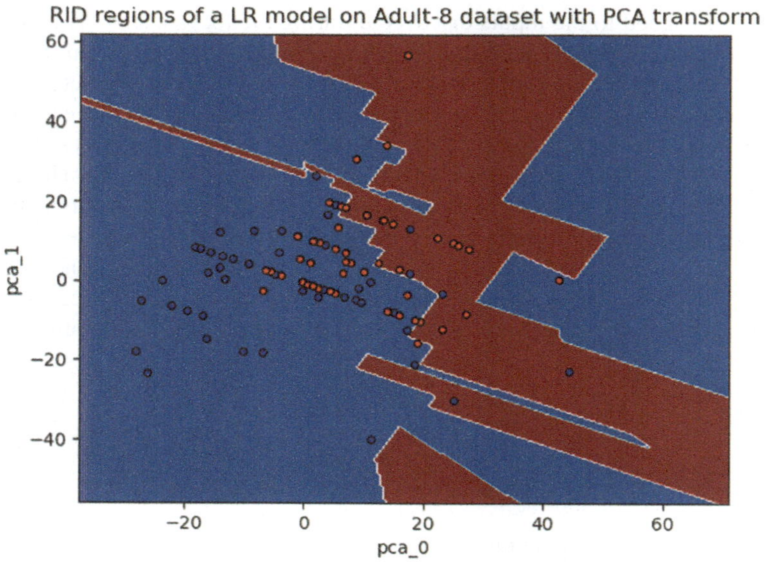

Fig. 2 Discriminatory input distribution for Logistic Regression model on Adult Income [26] dataset. The red area denotes the discrimination region and the figure shows that discriminatory inputs exist in regions and are not isolated

inputs. We use this observation to create an on-demand algorithm that judiciously generates synthetic inputs to determine the precise boundary between the discriminatory and non-discriminatory inputs. We illustrate the approach for such an intelligent generation of inputs to generate the RIDs.

4.1 RID

Given a model $M : X \rightarrow Y$ with a set of protected attributes P and an algorithm that can compute $D(x)$ (whether an input is discriminatory or not), our objective is to efficiently find a boolean function $R : X \rightarrow \{0, 1\}$ which approximates D, i.e., $\forall x \in X, R(X) = D(X)$, i.e., R is trying to simulate D which is defined for any input $x \in X$.

Our aim is to learn the function R using a model trained on a subset (*SeedX*) from the domain X such that it predicts well for the entire X. The challenge is to judiciously find the subset $SeedX$ from X. We address this challenge later in this section.

Interpretability. Note that we can use any binary classifier to learn R. However, that function is not necessarily going to be an interpretable one. For debugging, however, the interpretability of R is an essential criterion as one should understand what influences the separation of all discriminatory and non-discriminatory inputs and can also localize the discriminatory inputs in the input space.

As the first choice, we try to use a decision tree classifier, a well-known interpretable classifier, to classify discriminatory and non-discriminatory inputs. Essentially, a decision tree has many paths and each path is represented as the conjunction of predicates on features, where each predicate is of the form $v \geq c$, $v = c$, or $v < c$ where $v \in F$ and c is a constant value. *A region of Individual Discrimination or RID is defined as the conjunction of one or more such predicates on the non-protected features aiming to contain only the discriminatory inputs.* A RID *covers* an input $x \in X$ if x satisfies all its predicates. A RID contains all values of the protected features and therefore it is expressed without the protected features.

Our algorithm *RIDer* outputs a set of RIDs that aims to cover all the discriminatory inputs. The set of discriminatory inputs in the domain X defines the *True Discrimination Region* (i.e., $TDR = \{x \mid D(x) = 1\}$) and the set of RIDs corresponding to an R aims to accurately cover the TDR.

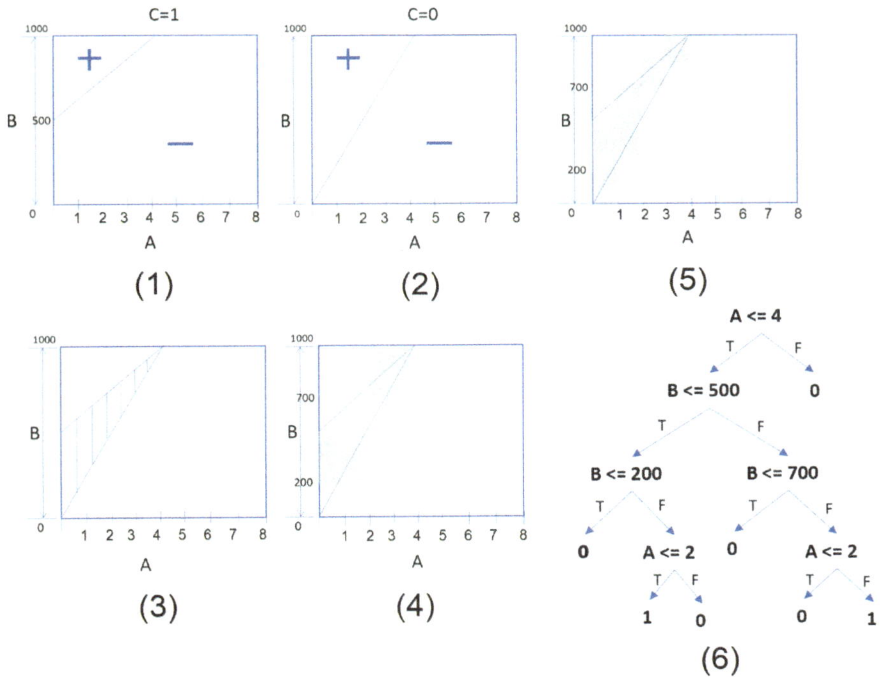

Fig. 3 **1** and **2** show the decision boundary for two different values of protected feature C. **3** shows the True discriminatory region (striped). **4** shows the RIDs (shaded box) found using a decision tree-based classifier shown in **6**. **5** shows the discriminatory region for an on-demand generation-based algorithm. For **1**–**5**, the horizontal axis shows values for feature A and the vertical axis show shows the values for feature B

Example. Let us consider an example where the set of features $F = \{A, B, C\}$ where C is a protected feature, i.e., $F = \{C\}$. A model M is trained to predict a binary class $Y = \{+, -\}$. Figure 3 shows M's decision boundary for two different values of the protected feature C in sub-figure (1) and (2). Essentially, we are showing a three-dimensional decision boundary in two-dimensional figures. It is evident from Sub-figures (1) and (2) that the decision boundary is different for different values of the protected feature. The striped region in (3) shows the TDR where for $Z = 1$, the model predicts "-" and, for $Z = 0$, the model predicts "+" which means that the inputs falling within this region are discriminatory based on the individual discrimination definition in [17]. The RID generation algorithm, described next, will try to approximate the striped region.

4.2 Determining RIDs

As stated earlier, our main objective is to obtain an interpretable RID with high precision and recall to predict if a sample is discriminatory.

It is evident from the above section that any RID generation algorithm can be formulated as a classification algorithm that aims to classify discriminatory and non-discriminatory samples. Moreover, the classifier also needs to be an interpretable one. An obvious choice for this classifier is a decision tree classifier.

We train a decision tree classifier with the input from Tr and label obtained through the function D signifying whether a sample is discriminatory (label 1) or not (label 0).[1] Once the decision tree is trained, RIDs can be obtained by traversing the root-to-leaf paths of the decision tree corresponding to label 1.

Example. Figure 3(6) shows the decision tree that generates two RIDs $A <= 2 \wedge 200 <= B \wedge B <= 500$ and $A >= 2 \wedge A <= 4 \wedge 700 <= B \wedge B <= 1000$ which are also shown as the two gray regions in Fig. 3(4). The two RIDs are obtained by forming the conjunction of the predicates for leaf value 1 signifying a discriminatory label.

Even though this simple algorithm addresses the RID generation problem, in the rest of this section, we propose an algorithm that optimizes the above decision tree-based "base" method. The first optimization is based on the metamorphic property of individual discrimination which essentially says that we do not need humans to provide the discriminatory label for any arbitrary sample.

[1] The choice of the training data depends on the debugging stage in Data and AI life cycle. An alternative choice at the model-building phase can be the union of training and validation data. Some debugging can be performed at runtime when several samples fail during runtime. At that point, training, validation, and test set for model M along with runtime workload can also be used.

4.3 On-Demand Sample Generation/Active Learning

Intuition Let us revisit the example given in Fig. 3(4). Consider the triangular TDR part in $(A = 1 \wedge B <= 200)$. Even though that region is discriminatory, it is not included in the RIDs as there were no discriminatory samples in the training data falling in that region. Though the base algorithm can perfectly distinguish the discriminatory samples present in the training data from the non-discriminatory ones, it is still going to give incorrect discrimination predictions for the samples in that triangular region.

Usually, in a model training scenario, the training procedure has to work with available training data. However, a crucial observation is that, here we want to replicate the individual discrimination behavior of an already trained model M. *Therefore, it is possible to generate more synthetic samples and find out whether the samples are discriminatory by the chosen discrimination check D. This follows from the metamorphic property of individual discrimination.*

Based on the above observation, we can now design an improved version of the base algorithm, essentially improving the choice of training data for the decision tree. The aim is to synthesize more samples and use the individual discrimination label from the check D while learning the decision tree. There are three important algorithmic design considerations as discussed below.

- When: One option is to generate the synthetic samples a-priori before starting the training process. This ensures that the remaining process does not change. The other choice is to generate the samples during the tree formation. This process can use information about the current decision tree and the regions that have already been identified. However, this changes the decision tree-building algorithm.
- Where: The next question is about deciding where in the input space, shall we generate the samples or which samples shall we generate. Note that the generation of such samples is aimed at obtaining a more accurate decision tree boundary.
- How many: Finally, how many samples we should generate is something we need to decide. If we generate too many samples, we can be very accurate in terms of approximating TDR boundary, but it may increase the overhead requiring a large number of splits.

Let us first describe the design choices that we have considered. Essentially, this algorithm does not generate the synthetic samples a-priori, rather it starts the decision tree-building algorithm by considering discriminatory labels on the training data. The decision tree generation algorithm tries splitting criteria on all attributes and finally, chooses to split on the attribute which results in maximum information gain. The feature along with its predicate form the node and the branches contain the outcome of the predicate signifying two regions. This process continues for each region and the data falling in that region. *The problem with this approach is that there are many samples to compute the split at the higher levels (near the root) of the tree. However, when the tree depth goes near the leaf, the number of samples in an impure (having samples with multiple labels) region becomes fewer and therefore, the split*

Algorithm 3: RID Generation algorithm

```
1  Function RIDer (SeedX, M, P, genthreshold, δ1, δ2)
2      isDiscriminatory = D(SeedX, M, P)
3      dtmodel = decision_tree(SeedX, isDiscriminatory,
        min_samples_leaf = genthreshold)
4      regions = []
5      foreach path, label, PathX ∈ pathsWExamples(dtmodel) do
6          if impure(PathX) then
7              genX = generate_samples(path, genthreshold - |PathX|)
8              pathX = pathX ∪ genX
9              regions1 = RIDer(pathX, M, p, genthreshold,δ1, δ2)
10             foreach path1 ∈ regions1 do
11                 path = path ∧ path1
12                 regions.add(path)
13             end
14         else
15             if label = 1 then
16                 regions.add(path)
17         end
18     end
19     return regions
20 end
21 Function D(SeedX, M, P)
22     disc′= disc=[]
23     disc_labels = []
24     foreach x ∈ SeedX do
25         if ∃x′, s.t. I(x, x′, M, P) holds then
26             disc_labels.add(1)
27             disc += x
28             disc′ += x′
29         else
30             disc_labels.add(0)
31         end
32     end
33     return disc_labels, disc, disc′
34 end
```

operation may not be accurate as it is determining the splitting boundary based on a small number of samples. This observation is noted from [11].

We use this observation to determine when and where synthetic samples need to be created. The algorithm *RID* is presented in Algorithm 3. The algorithm *RIDer*, along with the model M and the protected attributes P, takes $SeedX$ which is called using inputs in training data Tr, as in the base algorithm. $SeedX$ is then subjected to a discrimination testing procedure D (Line 2) to determine the discriminatory (class 1) and non-discriminatory samples (class 0) denoted by the variable $isDiscriminatory$. $SeedX$ along with $isDiscriminatory$ form the training data for the decision tree classifier. However, at Line 3 it stops decision tree formation early such that all impure nodes have samples lesser than a given threshold $genthreshold$ (another parameter to the algorithm) thereby finding sparse nodes where new samples can be generated. *RIDer* achieves this by using the $min\_samples\_leaf$ parameter in the Sckit-learn's decision tree implementation, thereby reusing all the optimizations of the existing implementation. Secondly, *RIDer* generates realistic synthetic samples to help create more precise decision boundaries at the sparse nodes. *RIDer*'s sample generation is motivated by a technique described in [30] which samples from the multivariate Gaussian distribution learned on the training data Tr. Specifically, *RIDer* ensures that the *realistic* synthetic samples (using method *generate_samples*) are generated in the impure regions resulting from the decision tree computation above. These regions correspond to the decision tree paths obtained by the function *pathWExamples* in Line 5. Note that all impure regions (at Line 7) will have a sample count lesser than $genthreshold$, otherwise, it would have been split by the decision tree construction at Line 3. The number of samples generated brings back the number of samples in that region equal to the $genthreshold$ by adding to the existing samples $PathX$ in that impure region (Lines 7–8). After such generation, *RIDer* is called recursively (Line 9) for each impure region. The algorithm terminates when all the generated decision tree regions (at Line 5) are pure.

In summary, *RIDer* uses on-demand generation of training samples wherever and whenever it is most required. Note that this process is an instance of active learning [33] which uses the D as an oracle and uses decision tree-specific knowledge for optimizing the oracle queries.

Example. Figure 3(5) shows the RIDs (shaded region) obtained using *RIDer*'s on-demand sample augmentation strategy. The algorithm essentially refines several regions including $(A <= 2 \wedge B <= 200)$ and $(2 <= A <= 4 \wedge 500 <= B <= 700)$ in Fig. 3(4) by generating samples in those regions.

5 Repairing

In this section, we describe the model repair algorithm which aims to repair individual discrimination in the model. We first present the repair setup. We then present the objective of the repair algorithm. Finally, we present an algorithm that performs a search strategy to minimize the objective.

5.1 Setup

At a high level, the Data and AI life cycle contains three phases. The first phase is the Data life cycle where data is pre-processed and made ready for building an AI model. The second phase is the model-building phase where data scientists try to produce the best model. The third phase is the post model-building phase, where the model is validated from a business perspective before it is deployed, monitored, and possibly retrained. In the second phase, the data scientist uses the validation data to iteratively strengthen or tune the model, and then one model is selected from multiple models based on their performance on the hold-out data. The first scenario where data scientists can use our repair technique is in the second phase where they have tested for individual discrimination with the training data and found discriminatory samples. At this point, they can use our method to generate a new training set (called the retraining data) and retrain the same model architecture with the new retraining data and ensure that it reduces the discrimination and does not affect the generalizability property of that model. The second scenario where our technique is useful is the post-deployment phase, where we see the discriminatory samples in the user-provided workload data which are served by the model.

To repair an existing model M, we assume to have access to the following information:

- Training Data (Tr): We assume that we have access to the training data. The training data will serve two purposes: (1) The repair algorithm will find the discriminatory samples within the training data. Note that, in principle, any seed data ($\subset X$) even without the labels could have been used to determine the discriminatory samples. (2) Our algorithm modifies the training data to create the retraining data.
- Model Architecture: We assume that we know the architecture of the model training algorithm of the model. For example, if M is a GradientBoosting Classifier model then this information is known to our analysis. This is available to the data scientists as described in the above scenario. Our algorithm, as described later, uses this information to train intermediate models to check the effectiveness of the changes.
- Invariant data: A set of instances of the training data and their corresponding model predictions for which the model does not want its prediction to change through retraining. The usefulness of this will be clear when we described the objective of the retraining next.

Our algorithm takes the training data and the model information and returns the retraining data and labels.

5.2 Objective

There are two specific objectives of repairing the model:

- Reduce the individual discrimination behavior of the model. We will measure such behavior for a given data as the percentage of the discriminatory data (also called the flip rate [3]) within it. This evaluation will be done on the training data and on the test/validation data. In summary, the objective is to minimize the flip rate.
- The generalization characteristics of the model should not get compromised. Specifically, the objective is to minimally change the predicted labels for a given set of data, which we call the invariant inputs. In the training phase, such invariant inputs can be the correctly predicted training data, i.e., the subset of the training

data for which the model predictions match with the ground truth labels. In the post-deployment phase, such invariant inputs can be the correctly predicated training data along with all the high-confidence predictions done so far for which the user has not flagged any error. The condition ensures that the effect of this change is minimal in the environment in which the model is used.

- The final minimization objective can be a weighted combination of these two minimization objectives. The user of the algorithm can decide on the corresponding weights.

We denote the invariant instances as I, where one possible realization of I, in the training phase, as described above could be $I = \{(x, M(x)) | M(x) == y, (x, y) \in Tr\}$. Let's say our algorithm generates the retraining data denoted as $RTr = (X_r, Y_r)$, and the retrained model trained on RTr is M_r. Therefore, for a given test input X and invariant instances I, the minimization objective is expressed as

$$\begin{aligned} \min_{M_r} \quad & f(X, P, M_r, I) \\ \text{s.t.} \quad & f(X, P, M_r, I) = \lambda_1.disc(X, P, M_r)/|X| \\ & + \lambda_2.|M_r(I_X) \neq I_Y|/|I| \end{aligned} \tag{2}$$

We want to minimize the objective without making any assumptions about the models M and M_r, since we want the analysis to be model-agnostic. Therefore, we can't make any assumption such as differentiability which can be used to effectively find a retrained model with minimum objective using techniques such as gradient descent. Next, we present a very simple iterative algorithm that does not guarantee minimality yet performs well in reducing the objective from the baseline procedures.

5.3 Iterative Algorithm

Our algorithm works on the principle that the individual discrimination is reduced when the model is trained with a discriminatory input x and its corresponding perturbed counterpart x' (i.e. $I(x, x', P, M)$ holds) are given the same label. This is essentially ensuring that the protected attribute has no effect on the prediction of x on the retrained model M_r. The first choice is therefore related to which label should be selected for the input x and x'. The two options here are $\hat{M}(x)$ and $\hat{M}(x')$. It also turns out that the effect of the inclusion of all discriminatory inputs along with their perturbed counterpart in the retraining set may not be optimal in minimizing the objective function. This raises another option—whether each discriminatory sample should be added to the retraining set.

Based on the above discussion, we formulate an algorithm that iteratively computes the effect of these three choices by incrementally adding each discriminatory input along with its perturbed counterpart. The algorithm is presented in Algorithm 4.

Algorithm 4: Iterative Repair algorithm

```
1  Function ItRepair (X, M, P, I)
2      disc, disc' = D(X, M, P)
3      cX = X − disc
4      cY = M̂(cX)
5      current_obj = obj(cX, cY, M, X, P, I)
6      foreach (x, x') ∈ zip(disc, disc') do
7          y = M̂(x), y' = M̂(x')
8          obj1 = obj(cX + x + x', cY + y + y, M, X, P, I)
9          obj2 = obj(cX + x + x', cY + y' + y', M, X, P, I)
10         if current_obj < obj1 ∧ current_obj < obj2 then
11             none
12         else if obj1 < obj2 then
13             cX = cX + x + x'
14             cY = cY + y + y
15             current_obj= obj1
16         else
17             cX = cX + x + x'
18             cY = cY + y' + y'
19             current_obj= obj2
20         end
21     end
22     return cX, cY
23 end
24 Function D (X, M, P)
25     disc'= disc=[]
26     foreach x ∈ X do
27         if ∃x', s.t. I(x, x', M, P) holds then
28             disc += x
29             disc' += x'
30     end
31     return disc, disc'
32 end
33 Function obj (X_r, Y_r, M, X, P, I)
34     M_r = M.clone()
35     M_r.fit(X_r,Y_r)
36     return f(X,P,M_r,I)
37 end
```

The algorithm takes a set of inputs (X) which is called with the training data inputs ($Tr.X$) if the repair algorithm is used in the training phase of the life cycle, along with the trained model M the set of protected attributes P and invariant instances I. The algorithm first computes the discriminatory inputs ($disc$) and their perturbed counterpart $disc'$ (Line 2). Then it computes the non-discriminatory inputs which form the current retraining inputs (cX) and computes the objective with the retraining data. At every iteration it tries to see the effect of the addition of one discriminatory input (x) and its perturbed counterpart x' with the same labels—either x's label (Line 8) or x''s label (Line 9) from the original model M. It updates the current retraining set if the addition results in a reduction of the objective value. Note that computation of the objective value requires retraining a model with the same architecture obtained using the clone function in Line 36 (as in Scikit-Learn).

6 Experimental Results

In this section, we show preliminary experimental results related to individual discrimination testing, debugging, and repair.

6.1 Benchmarks

We have assessed the effectiveness of our approach on two well-known open-source fairness datasets [13] as listed in Table 1. Note that these datasets are used by prior fairness testing papers as well. We consider a benchmark as the combination of a dataset and a trained model. We choose 3 models per dataset. We have experimented by considering gender-related attributes in these benchmarks as the protected attribute.

6.2 Setup and Configuration

All the algorithms are implemented as ∼4000 lines of code in Python. All the experiments are performed in a machine running macOS 10.14, having 16 GB RAM, 2.7 GHz CPU running Intel Core i7 running Python 3.7. For every dataset, we have trained a varied set of target models (with test accuracy as reported in Table 1) using default configuration as in *Scikit-learn* [29] with 80:20 as the train-test (Tr:Te) split ratio. We use a simple discrimination function (I) as used in THEMIS [17], i.e.,

Table 1 Benchmarks with varied model types and training data sizes. LR-Logistic Regression, KNN-k-nearest Neighbors, MLP-Multilayer Perceptrons

Dataset (Prot. Attr)	Train size	Model	Test accuracy (%)
Adult-8 [26] (sex)	26048	LR	77.18
		MLP	78.75
		KNN	78.27
Car rentals [3] (Gender)	388	LR	71.42
		MLP	77.55
		KNN	74.49

perturbing the protected attribute in the sample, and marking it as discriminatory if the model's prediction changes. This implementation choice is motivated by what is followed by all the existing implementations of individual discrimination testing algorithms.

6.3 Research Questions

Following are the high-level research questions, based on which we have designed our experiments. Note that, these are selected experiments in order to demonstrate the effectiveness of our approach and do not include various possible comparisons, detailed statistics, and ablation studies.

- Test. How effective is the symbolic test generation algorithm in terms of generating discriminatory test cases?
- Debug. How effective is the RIDer algorithm for predicting discriminatory samples?
- Repair. How effective is the repair algorithm to repair individual discrimination while still maintaining the generalizability of the models?

6.4 Metrics

- Flip-rate is defined as the percentage of actual discriminatory samples in the generated test set. This metric is used in Test and Repair.
- Given the test data t, RID precision is the percentage of samples in t, covered by RIDs or predicted as discriminatory, that is actually discriminatory.
- Given the test data t, RID recall is the percentage of discriminatory samples, from the set t, that are covered by the RIDs or truly identified as discriminatory. Both RID precision and recall are used in Debug.

6.5 Results

6.5.1 Test

The effectiveness of our symbolic-execution-based test generation algorithm, called SG, is shown in Table 2. The flip rate is shown in two different stages of the algorithm. The global stage performs a clustering-based algorithm and then performs symbolic execution on the decision tree to generate test cases. The local stage perturbs the discriminatory samples found in the global stage.

Table 2 Effectiveness of test generation (SG) in finding individual discrimination in the models

Dataset	Model	Global flip rate	Local flip rate
Adult-8	LR	10.00	65.8
	MLP	5.10	42
	KNN	4.10	25.3
Car	LR	50.30	66.3
	MLP	45.27	33
	KNN	51.20	42.1

Conclusion. The average flip rates across 6 benchmarks are 27.6% and 45.75% for the two phases which demonstrates that our test case generation is effective. For more results, please refer to [3].

6.5.2 Debug

We compare the effectiveness of the RID generation algorithms. The first version, RIDerNaive, is based on learning a decision tree to classify discriminatory and non-discriminatory samples. The second algorithm, RIDerGen, is where more samples are generated on-demand to compute the decision tree as described in Algorithm 3. All the RID generation algorithms are trained on Tr and tested on Te.

The result is shown in Table 3. *RIDerGen* achieve better precision (13.62%) on average than *RIDerNaive*. A similar trend is seen for recall with *RIDerGen* achieving on average 13.62% better recall than *RIDerNaive*.

Conclusion. Sample generation in *RIDerGen* is effective in increasing the accuracy as compared to the naive decision tree-based algorithm.

Table 3 Precision and recall comparison between RIDerNaive and RIDerGen algorithms

Dataset	Model	Naive		Gen	
		Pr(%)	Re(%)	Pr(%)	Re(%)
Adult-8	LR	92.96	91.27	95.71	94.15
	MLP	94.35	94.16	96.49	95.25
	KNN	60.49	59.27	68.97	60.48
Car	LR	85.71	100	100	100
	MLP	100	100	100	100
	KNN	100	100	100	100

Table 4 Comparison of flip rate and accuracy before and after repair on training and test data

Dataset	Model	Flip-rate				Accuracy			
		Train		Test		Train		Test	
		Org	Rep	Org	Rep	Org	Rep	Org	Rep
Adult-8	LR	15.98	0.97	16	0.91	0.77	0.76	0.78	0.76
	MLP	15.78	1.49	15.51	1.66	0.79	0.77	0.79	0.77
	KNN	6.08	2.2	7.62	4.08	0.78	0.78	0.87	0.85
Car	LR	24.23	0.52	24.49	0	0.71	0.7	0.72	0.71
	MLP	27.84	1.55	24.49	2.04	0.78	0.79	0.91	0.88
	KNN	26.03	12.11	29.59	11.22	0.74	0.74	0.91	0.88

6.5.3 Repair

The goal of the experiment is to compare two metrics (1) the flip rate and (2) the accuracy of the model before and after repair.

The experiment is performed by selecting all discriminatory samples in the training data as repair data across all benchmarks. The result is shown in Table 4 where *org* refers to the result before repairing the model and *Rep* after repairing. The result shows on average X% decrease in flip rate across all benchmarks. For the test data, the flip rate decrease is 82% and 70% for training and test data respectively across a1 1 benchmarks with 4 benchmarks showing more than a 90% decrease. The average accuracy loss for train and test data is 0.6% and 2.5% for all the models.

Conclusion. Our repair algorithm is able to achieve a substantial decrease in the discrimination for the majority of benchmarks without losing accuracy.

7 Related Works

This section describes the literature related to individual discrimination testing, debugging, and repair.

Individual Discrimination Testing. Galhotra et al. [17] introduced the notion of fairness testing in software engineering research. Their tool, called THEMIS, generates random data and perturbs only the protected attributes, and checks if the predictions differ. AEQUITAS [37] uses a two-phase approach: in the global phase, it generates random samples and checks for discrimination and then in the next (local) phase, searches in the neighborhood of already identified discriminatory samples for more discrimination. Our test case generation algorithm is first published as the SG algorithm [3] that uses diversified search using clustering in the global phase with the help of local explanation and then, performs local search near discriminatory samples by perturbing less-influential non-protected attributes. Handcrafted tests are used in

FairTest [36] to quantify four different discrimination scores. Zhang et al. [38] take a fully transparent, white-box gradient-based perturbation approach to guide the search for discriminatory samples.

Individual Fairness Debugging. We have not come across any work which tries to perform localization of individual discrimination. However, there are some notable works related to model debugging. SliceFinder [9] tries to identify subsets or cohorts in the training data where the model performs *poorly*. Similar to ours, they try to find interpretable slices and use decision tree-based solutions to distinguish between passing and failure samples. However, they primarily use the algorithm to debug accuracy (e.g., class label mismatch with respect to ground truth) failures. Since accuracy is not a metamorphic property, SliceFinder algorithm resembles our naive decision tree-based algorithm. SliceFinder claims to be generic for any property which can be defined using a scoring function (also used for debugging group discrimination metric called equalized odds). Concurrent with our work, the MD algorithm ([10]) tries to explain the mispredicted samples of ML models using rule sets. Unlike [9], which optimizes for accuracy, MD optimizes for the precision of the rule sets and performs a grid search to obtain rules with high precision for a given threshold of recall. As shown by our experiments, the MD algorithm does not perform well even in comparison to the naive decision tree-based algorithm in terms of precision and rule size. Compared to the flexible path conditions generated by the decision tree, the MD algorithm a-priori creates a fixed set of predicates to form the rules. This is the reason for less precision.

Individual Discrimination Repair. Note that there exists many in-processing, preprocessing, and post processing algorithms for mitigating group discrimination [2, 31]. CAPUCHIN [31] performs repair of training data by insertions and deletions of database tuples, however, they do not consider model predictions or data regions where a model is unfair. Moreover, their notion of fairness, interventional fairness, is based on causal dependency and group fairness. There are only a few works on repairing individual discrimination [8, 37]. AEQUITAS [37] uses randomly sampled data augmentation technique for repairing and [8] repairs group discrimination and individual discrimination with specific kinds of changes to repair data imbalance and existing data label and does not consider the failure samples to repair. Their technique is orthogonal to ours.

8 Conclusion

In this book chapter, we present solutions to three important problems in the AI life cycle related to individual discrimination behavior observed in machine learning models. Our algorithm [3] was one of the earliest methods for test case generation for checking individual discrimination for black-box machine learning models. The algorithm uses a local explainer to judiciously perturb the attributes in order to generate more failures, i.e., samples exhibiting individual discrimination. Once such failure test cases are obtained, our RIDer algorithm localizes the discriminatory samples in interpretable regions. Such localization helps developers and downstream users to understand the correlation between feature values and the discriminatory behavior of the model. Finally, we present a repair algorithm to repair individual discrimination without compromising on accuracy.

A major limitation of our work is that it is applicable only to tabular modality. Future work should focus on extending individual discrimination testing to time series, text, speech, and images. Debugging and repairing individual discrimination has still not been investigated thoroughly in the literature, and we believe our work will start many future avenues to improve our results. We believe that we can use RIDs to efficiently repair the model and further improve the precision and recall of RIDs by introducing non-oblique decision boundaries. Future work should also focus on extending debugging and repairing algorithms to other data modalities. Creating interpretable regions is a major challenge for text modalities, and therefore, we should focus on alternate ways of finding the relationship between discriminatory behavior and text inputs.

References

1. IBM Watson Studio - AutoAI, Last accessed 15th Oct 2020
2. IBM AIF360, Last accessed 19th Feb 2021
3. Aggarwal A, Lohia P, Nagar S, Dey K, Saha D (2019) Black box fairness testing of machine learning models. ESEC/FSE
4. Basu K, Basu T, Buckmire R, Lal N (2019) Predictive models of student college commitment decisions using machine learning. Data 4:65, 05
5. Binns R (2020) On the apparent conflict between individual and group fairness. In: Proceedings of the 2020 conference on fairness, accountability, and transparency, pp 514–524
6. Breiman L, Friedman J, Stone CJ, Olshen RA (1984) Classification and regression trees. CRC Press
7. Cadar C, Ganesh V, Pawlowski PM, Dill DL, Engler DR (2023) Exe: automatically generating inputs of death. In: CCS '06, pp 322–335
8. Chakraborty J, Majumder S, Menzies T (2021) Bias in machine learning software: why? how? what to do? Association for Computing Machinery, New York, NY, USA
9. Chung Y, Kraska T, Polyzotis N, Tae KH, Whang SE (2019) Automated data slicing for model validation: a big data-ai integration approach. IEEE Trans Knowl Data Eng 32(12):2284–2296
10. Cito J, Dillig I, Kim S, Murali V, Chandra S (2021) Explaining mispredictions of machine learning models using rule induction. In: Proceedings of the 29th ACM joint meeting on European software engineering conference and symposium on the foundations of software engineering, pp 716–727
11. Craven MW (1996) Extracting comprehensible models from trained neural networks. PhD thesis. AAI9700774
12. Dressel J, Farid H (2018) The accuracy, fairness, and limits of predicting recidivism. Sci Adv 4:eaao5580
13. Dua D, Graff C (2017) UCI machine learning repository
14. Dwork C, Hardt M, Pitassi T, Reingold O, Zemel R (2012) Fairness through awareness. In: ITCS 2012, pp 214–226

15. Esteva Andre, Robicquet Alexandre, Ramsundar Bharath, Kuleshov Volodymyr, DePristo Mark, Chou Katherine, Cui Claire, Corrado Greg, Thrun Sebastian, Dean Jeff (2019) A guide to deep learning in healthcare. Nat Med 25(1):24–29
16. Feldman M, Friedler SA, Moeller J, Scheidegger C, Venkatasubramanian S (2015) Certifying and removing disparate impact. In: proceedings of the 21th ACM SIGKDD international conference on knowledge discovery and data mining, pp 259–268
17. Galhotra S, Brun Y, Meliou A (2017) Fairness testing: testing software for discrimination. In: ESEC/FSE. ACM, New York, NY, USA, pp 498–510
18. Godefroid P (2007) Compositional dynamic test generation. In: Proceedings of the 34th annual ACM SIGPLAN-SIGACT symposium on principles of programming languages, POPL '07. ACM, New York, NY, USA, pp 47–54
19. Godefroid P, Klarlund N, Sen K (2023) Dart: directed automated random testing. In: PLDI' 05, pp 213–223
20. Huang B, Kechadi MT, Buckley B (2012) Customer churn prediction in telecommunications. Expert Syst with Appl 39(1):1414–1425
21. Joseph M, Kearns M, Morgenstern JH, Roth A (2016) Fairness in learning: classic and contextual bandits. In: Lee D, Sugiyama M, Luxburg V, Guyon I, Garnett R (eds) Advances in neural information processing systems, vol 29. Curran Associates, Inc.
22. Kusner M, Loftus J, Russell C, Silva R (2017) Counterfactual fairness. In: NIPS. Curran Associates Inc, USA, pp 4069–4079
23. Lahoti P, Weikum G, Gummadi K (2019) ifair: learning individually fair data representations for algorithmic decision making. In: 2019 IEEE 35th international conference on data engineering (ICDE), pp 1334–1345
24. Le TT, Fu W, Moore JH (2020) Scaling tree-based automated machine learning to biomedical big data with a feature set selector. Bioinformatics 36(1):250–256
25. Mehrabi N, Morstatter F, Saxena N, Lerman K, Galstyan A (2019) A survey on bias and fairness in machine learning. arXiv:1908.09635
26. Mothilal RK, Sharma A, Tan C (2020) Explaining machine learning classifiers through diverse counterfactual explanations. FAT* '20
27. Mukerjee Amitabha, Biswas Rita, Deb Kalyanmoy, Mathur Amrit P (2002) Multi-objective evolutionary algorithms for the risk-return trade-off in bank loan management. ITOR 9:583–597
28. Pedregosa F, Varoquaux G, Gramfort A, Michel V, Thirion B, Grisel O, Blondel M, Prettenhofer P, Weiss R, Dubourg V, Vanderplas J, Passos A, Cournapeau D, Brucher M, Perrot M, Duchesnay E (2011) Scikit-learn: machine learning in python. J Mach Learn Res 12:2825–2830
29. Pedregosa F, Varoquaux G, Gramfort A, Michel V, Thirion B, Grisel O, Blondel M, Prettenhofer P, Weiss R, Dubourg V et al (2011) Scikit-learn: machine learning in python. J Mach Learn Res 12:2825–2830
30. Ribeiro MT, Singh S, Guestrin C (2016) Why should i trust you?: explaining the predictions of any classifier. In: Proceedings of the 22nd ACM SIGKDD international conference on knowledge discovery and data mining. ACM, pp 1135–1144
31. Salimi B, Rodriguez L, Howe B, Suciu D (2019) Interventional fairness: causal database repair for algorithmic fairness. In: Proceedings of the 2019 international conference on management of data, pp 793–810
32. Sen K, Marinov D, Agha G (2005) Cute: a concolic unit testing engine for c. In: Proceedings of the 10th European software engineering conference held jointly with 13th ACM SIGSOFT international symposium on foundations of software engineering, ESEC/FSE-13. ACM, New York, NY, USA, pp 263–272
33. Settles B (2009) Active learning literature survey
34. Sun Y, Wu M, Ruan W, Huang X, Kwiatkowska M, Kroening D (2018) Concolic testing for deep neural networks
35. Szegedy C, Zaremba W, Sutskever I, Bruna J, Erhan D, Goodfellow IJ, Fergus R (2013) Intriguing properties of neural networks. CoRR, abs/1312.6199

36. Tramèr F, Atlidakis V, Geambasu R, Hsu D, Hubaux J, Humbert M, Juels A, Lin H (2017) Fairtest: discovering unwarranted associations in data-driven applications. In: 2017 IEEE European symposium on security and privacy (EuroS P), pp 401–416
37. Udeshi S, Arora P, Chattopadhyay S (2018) Automated directed fairness testing. ASE
38. Zhang P, Wang J, Sun J, Dong G, Wang X, Wang X, Dong JS, Dai T (2020) White-box fairness testing through adversarial sampling. In: Proceedings of the ACM/IEEE 42nd international conference on software engineering, pp 949–960, 2020

Group and Individual Fairness in Clustering Algorithms

Shivam Gupta, Shweta Jain, Ganesh Ghalme, Narayanan C. Krishnan, and Nandyala Hemachandra

Abstract Clustering is a classical unsupervised machine learning technique. It has various applications in criminal justice, automated resume processing, bank loan approvals, recommender systems, and many more. Despite being so popular, traditional clustering algorithms may result in discriminatory behavior towards a group of people (or individuals) and have societal impacts. It has led to the study of fair clustering algorithms that aim to minimize the clustering cost while ensuring fairness. This chapter outlines existing group and individual fairness notions, discusses their relationships, and comprehensively categorizes the current algorithms. The chapter further discusses the advantages and disadvantages of existing algorithms in terms of theoretical guarantees, time complexity, and reproducibility. Finally, the chapter concludes with a discussion of new directions and open problems in the field of fair clustering.

Keywords Clustering · Algorithmic fairness · Group fairness · Individual fairness · Linear program · Optimization

---

S. Gupta · S. Jain
Indian Institute of Technology Ropar, Bara Phool, India
e-mail: shivam.20csz0004@iitrpr.ac.in

S. Jain
e-mail: shwetajain@iitrpr.ac.in

G. Ghalme
Indian Institute of Technology Hyderabad, Kandi, India
e-mail: ganeshghalme@ai.iith.ac.in

N. C. Krishnan (✉)
Indian Institute of Technology Palakkad, Kanjikode, India
e-mail: ckn@iitpkd.ac.in

N. Hemachandra
Indian Institute of Technology Bombay, Mumbai, India
e-mail: nh@iitb.ac.in

A. Mukherjee et al. (eds.), *Ethics in Artificial Intelligence: Bias, Fairness and Beyond*, Studies in Computational Intelligence 1123,
https://doi.org/10.1007/978-981-99-7184-8_2

1 Introduction

Increasing adoption of machine learning (ML) models for real-world applications, such as self-driving cars, college admissions, algorithmic pricing, data summarization, and recidivism, has exposed the prejudiced outlook [33, 72] towards individuals and groups characterized through protected attributes such as race and gender. Current model training practices do not explicitly counter the dataset biases resulting in machine learning models reflecting these biases. Further, undesirable behavior can result from the model using protected information for the predictions (*algorithmic bias*) [10, 65, 66]. Figure 1 illustrates a few anecdotes observed in recent years. Thus, designing fair and accurate machine learning models is central to improving the models' trustworthiness [38].

Applications such as automated resume processing [31, 45, 57], loan application screening, data summarization [49, 55, 81], and criminal risk prediction [50] are driving the recent efforts to develop fair clustering algorithms [4, 22, 47]. The prevalence of anthropological factors such as discrimination based on gender, race, and ethnicity in the data has resulted in a plethora of techniques to achieve group fairness. Group fairness demands that different groups should be treated in an unbiased manner. For instance, discrimination, such as shortlisting fewer qualified females for high-paying jobs, is unfair to the female group. In the clustering context, group fairness techniques aim to lower the clustering cost while achieving approximately equal representation for all groups.

Group fairness does not ensure fair treatment for a particular individual. The trait of human envy might still make an individual discontented. For example, an employee might feel discriminated against or left out if similar employees receive a favorable appraisal. There are algorithms in the literature that guarantee individual fairness

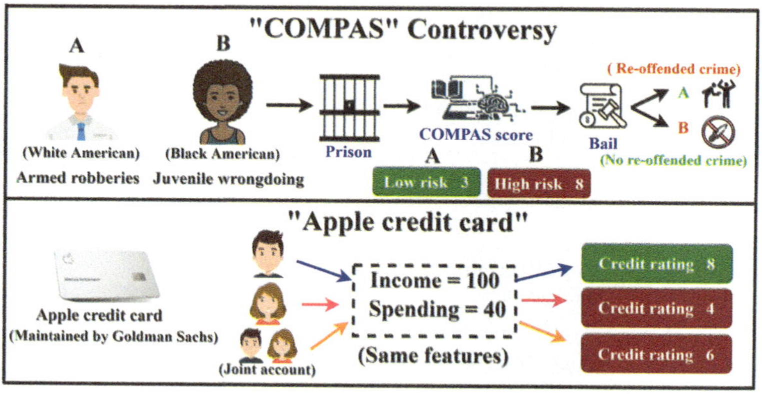

Fig. 1 Controversies that necessitate fair machine learning—**a** U.S. Criminal risk prediction system COMPAS opposed black Americans by tagging them risky [51]. **b** Apple's algorithm rated females with low credit scores in individual and joint accounts (with males) irrespective of similar feature values [69]

Fig. 2 Real-world scenarios **a** Group fairness—resume shortlisting systems use clustering [45, 57] to help HR quickly scrutinize cluster heads. A company allocates its employee among new branches while maintaining diversity. **b** Individual fairness—in marketing, similar individuals availing different offers feel discontented. Likewise, an insurance company providing similar health coverage satisfy clients better

[54, 62]. While group and individual fairness are the most popular, other types of fairness have been studied in the clustering literature. These include proportional fairness [68], core fairness [59], social fairness [30, 37, 40, 63], colorful fairness [7, 8, 15, 48], representative fairness [1, 71], and community fairness [21]. This chapter primarily focuses on the widely studied group and individual fairness (also referred to as equity and equitable fairness, respectively) as they have many real-world implications (refer Fig. 2a and b respectively).

Typical fair clustering solutions aim to minimize the standard clustering objectives while simultaneously enforcing fairness. The stage (*pre-processing, in-processing, and post-processing*) at which fairness constraints are enforced is a key differentiating factor of the existing algorithms. Pre-processing techniques alleviate data bias before clustering by adding restrictions on the distribution of points among clusters by a clustering algorithm [11, 29]. In contrast, in-processing techniques intertwine the clustering and fairness imposition parts of the algorithm [3, 41]. Finally, fairness is an afterthought for post-processing interventions that typically redistribute the instances of clusters obtained by vanilla clustering to obtain fair clusters [17, 42].

The solution framework is another distinguishing factor for fair clustering techniques. While some approaches add a regularizer term encoding the fairness to the clustering objective cost [2, 85], other approaches propose linear programming (LP) formulation of the fair clustering problem with linear fairness constraints [3, 17, 34, 42]. Other group fair clustering solutions include tree-based structures, round-robin allocation, and decision problems (such as min-cost flow and perfect matching algorithms) [11, 20, 29, 41, 61]. Solution approaches to individual fairness clustering include LP-based formulations [54, 70], local search approaches [62], dynamic programming [56], and combinatorial optimization [82]. Note that all the existing algorithms apply only to certain clustering objectives (k-means/k-median/k-center). So applicability of the approach is another differentiating factor. Figure 3 shows the taxonomy of existing algorithms categorized along (i) different stages of implementation, (ii) underlying solution frameworks, and (iii) applicability.

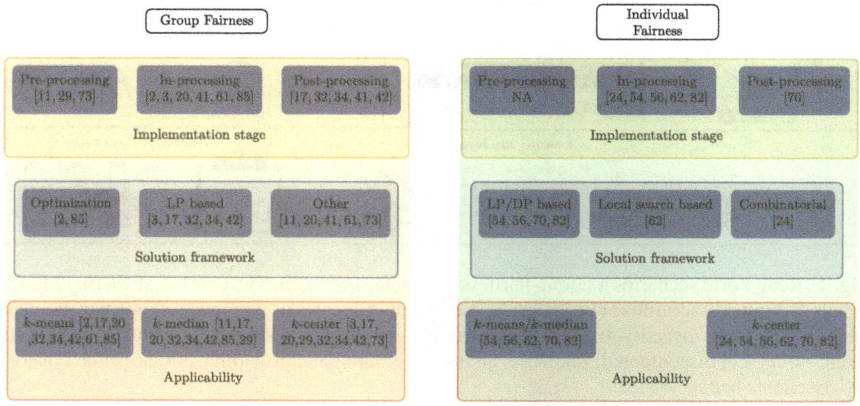

Fig. 3 Taxonomy of group and individual fairness in clustering algorithms

Many state-of-the-art (SOTA) techniques are supported by theoretical guarantees on fairness and the quality of the clusters, along with detailed cost approximation and computational complexity analysis. With the growing number of new fairness notions and algorithms, a comprehensive review that goes beyond summarizing the algorithms [27] to discuss the theoretical underpinnings, computational challenges, and relationships between different fairness notions will help in identifying the successes and future directions. While it has been shown that group and individual fairness may not exist together [6, 32], we discover interesting correlations between prevalent fairness notions from the experiments on two real-world datasets. Specifically, we observe that satisfying one kind of fairness has a positive inductive effect on another kind of fairness. Overall, there is a compelling need to comprehensively discuss the seemingly disparate fair clustering literature along the group and individual fairness dimensions. This is the primary objective of the current chapter.

2 Preliminaries

Clustering deals with partitioning a set $X \subseteq \mathbb{R}^h$ of n points into $[k] = \{1, 2, \ldots, k\}$ clusters, to obtain a clustering $C = \{C_1, C_2, \ldots, C_k\}$. Each cluster is represented by a center and let C represent the set of centers $= \{c_1, c_2, \ldots, c_k\}$. Let $d : X \times X \rightarrow \mathbb{R}^+$, denote a distance function obeying triangular inequality that measures the similarity between any two points. The goal of the algorithm is to minimize the following:

Definition 1.1 (*Objective Cost*) Given X and an assignment function $\phi : X \rightarrow C$ that assigns each point $x \in X$ to a center. The objective cost of clustering is

$$L_p(X, \phi) = \left(\sum_x d(x, \phi(x))^p \right)^{1/p} \tag{1}$$

The cases with $p = \{1, 2, \infty\}$ norms represent standard k-median, k-means, and k-center vanilla clustering respectively. It is well known that clustering is an NP-Hard problem. Let ρ be the cost approximation factor and T be the time complexity of the vanilla clustering problem [23, 39, 44, 60, 78].

Fair clustering algorithms, while minimizing objective cost, also ensure fair clusters. Binary or multi-value protected (or sensitive) attribute such as gender and race is central to the ***Group level fairness*** that arises from unintentional discrimination based on these attributes. Let us denote the value of the protected attribute for a point x by $S(x)$ and set of all protected attribute values by S. The protected attribute usually corresponds to the disadvantaged groups and is known apriori to algorithms as additional input. ***Individual level fairness*** is not tied to protected attributes but rather to 'similar' individuals. Let each point x identify itself with other similar points represented by the set S_x. Further, let $r(x)$ be the minimum radius of a ball $\mathcal{B}(x, r(x))$ centered around x that contains n/k points, where k is the number of clusters. We now define different notions of fairness under each level of fairness.

3 Group Fairness

The notion of group fairness is motivated by the principle of *disparate impact* [19, 33] from the classification literature. Group fairness in clustering guarantees minimum representation of each protected group in every cluster. Most of the current literature focuses on achieving group fairness against a single protected attribute, aka non-overlapping group identities. However, in practice, the group identities often overlap. For example, a person can belong to two protected attributes: race as black and gender as female. Overlapping identities are the focus of a few recent developments [17, 42] (discussed in detail later in Sect. 6.3). The underlying theme of prevalent group fairness notions is maintaining diversity in each cluster.

Definition 1.2 (τ-BALANCE) For a binary valued protected attribute $S = \{a, b\}$ and a scalar $\tau \in [0, 1]$, a clustering C is said to be τ-BALANCE ([29]) if for all $C_j \in C$

$$\min\left(\frac{\sum_{x\in C_j}\mathbb{I}(S(x)=a)}{\sum_{x\in C_j}\mathbb{I}(S(x)=b)}, \frac{\sum_{x\in C_j}\mathbb{I}(S(x)=b)}{\sum_{x\in C_j}\mathbb{I}(S(x)=a)}\right) \geq \tau. \quad (2)$$

The τ-BALANCE problem is formulated as *max-min* problem, which maximizes the minimum balance of a clustering. This notion does not allow the user to provide a trade-off between the clustering objective and fairness. Further, it is easy to see that the maximum value of τ in τ-BALANCE is equal to the *dataset ratio*, i.e., the setting when each cluster exhibits an equal number of points from every group.

Let us take an example to understand the notion with the binary-protected attribute taking two values, red and blue. A clustering algorithm divides the points into two clusters with 12 red and 3 blue points in one cluster; and 3 red and 3 blue in another cluster. Such a clustering is said to obey 0.25-Balance i.e. $\min\left(\min(\frac{12}{3}, \frac{3}{12}), \min(\frac{3}{3}, \frac{3}{3})\right)$. The dataset ratio, however, is $\frac{6}{15} = 0.4$, and as can be seen, red points significantly dominate blue points in the first cluster.

τ-BALANCE though captures the notion of having an almost equal number of points from different groups, it is still restricted to binary protected attribute. More generic fairness notions, i.e., Restricted dominance ($\boldsymbol{\tau}$-RD) and Minority protection ($\boldsymbol{\tau}$-MP), avoid dominance and preserve the minimal representation of a single group, respectively are defined below:

Definition 1.3 ($\boldsymbol{\tau}$-RD) A clustering C is said to obey restricted dominance with respect to $\boldsymbol{\tau}$ (i.e. $\boldsymbol{\tau}$-RD [17]) if for all $a \in S, C_j \in C$,

$$\sum_{x\in C_j}\mathbb{I}(S(x)=a) \leq \tau_a|C_j|. \quad (3)$$

Definition 1.4 ($\boldsymbol{\tau}$-MP) A clustering C is said to obey minority protection with respect to $\boldsymbol{\tau}$ (i.e. $\boldsymbol{\tau}$-MP [17]) if for all $a \in S, C_j \in C$,

$$\sum_{x\in C_j}\mathbb{I}(S(x)=a) \geq \tau_a|C_j|. \quad (4)$$

In the same example as above, if we consider notions of $\boldsymbol{\tau}$-MP and $\boldsymbol{\tau}$-RD, then the clustering has a lower bound of (0.2, 0.2)-MP and upper bound of (0.8, 0.8)-RD.

All these notions further determine fairness concerning the points with different protected attribute values but within a cluster. The $\boldsymbol{\tau}$-FAIR notion restricts the number of points from each group within a cluster concerning the total number of points from the same group in the whole dataset.

Definition 1.5 ($\boldsymbol{\tau}$-FAIR) A clustering C is said to obey $\boldsymbol{\tau}$-ratio fairness with respect to $\boldsymbol{\tau}$ (i.e.$\boldsymbol{\tau}$-FAIR [41]) if for all $a \in S, C_j \in C$,

$$\sum_{x \in C_j} \mathbb{I}(S(x) = a) \geq \tau_a \sum_{x \in X} \mathbb{I}(S(x) = a). \tag{5}$$

Similarly, as above, it is easy to see that this clustering will be (0.2, 0.5)-Fair.[1] Next, we define the τ-FE notion. It also considers fairness with respect to the number of points in each cluster but leads to a continuous and convex optimization objective. f-divergence can also be used instead of KL-divergence [6].

Definition 1.6 *(τ-FE)* The fairness error (τ-FE [85]) of a clustering C with respect to a given vector $\boldsymbol{\tau}$ is defined as:

$$\sum_{j \in [k]} \mathcal{D}_{KL}(\boldsymbol{\tau} || P_j) = \sum_{j \in [k]} \sum_{a \in S} -\tau_a \log P_j^a \tag{6}$$

where, $\mathcal{D}_{KL}$ is the Kullback-Leibler (KL) divergence and P_j^a is the fraction of points with protected group value a in cluster j.

4 Individual Fairness

The fundamental principle behind individual fairness is that similar individuals expect similar treatment. Any deviation would induce an unfair feeling in an individual [52, 77]. Various notions of individual fairness differ in how individuals perceive similarity.

Definition 1.7 (α-FR *Fairness*) A clustering C is said to be α-FR fair [62] if for $\alpha \geq 0$, C obeys

$$d(x, \phi(x)) \leq \alpha.r(x) \quad \forall x \in X. \tag{7}$$

The α-FR notion assures that any point x has its center within a radius containing n/k neighbors of x. This notion is restrictive as the neighbors of x are also determined using distance function $d(.)$. We now define more generalized notions (Fig. 4).

[1] $\boldsymbol{\tau}$ vector is written in the form (red, blue) respectively in $\boldsymbol{\tau}$-MP, $\boldsymbol{\tau}$-RD and $\boldsymbol{\tau}$-FAIR notion.

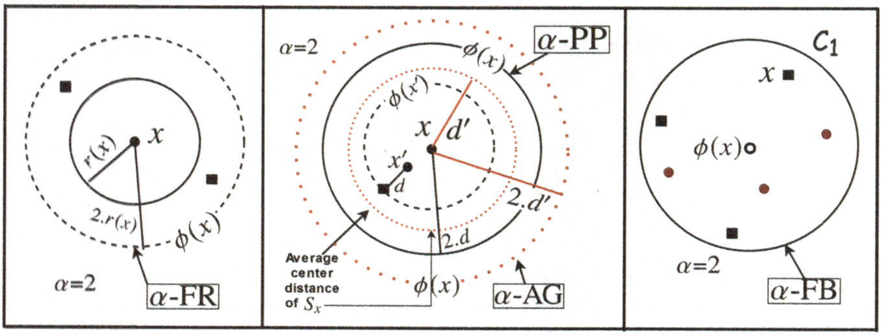

Fig. 4 Individually fair notions: **a** Given a point x, 2-FR demands that center for x (denoted by $\phi(x)$) lies at most within $2r(x)$ from x (dotted line). **b** 2-PP suggests the center be within twice the minimum center distance of a point, say x' in similarity set S_x, i.e., within $2d$. 2-AG relaxes the distance to $2d'$ by taking the average distance d'. **c** 2-FB demands at least two points of similar type in the cluster

Definition 1.8 *(α-PP Equitable Fairness)* (Per point Fairness) [24] A clustering C is said to be α-PP fair if for $\alpha \geq 0$, C obeys

$$d(x, \phi(x)) \leq \alpha . \min_{x' \in S_x} d(x', \phi(x')) \quad \forall x \in X \tag{8}$$

Definition 1.9 *(α-AG Equitable)* (Aggregate Fairness) [24] A clustering C is said to be α-AG fair if for $\alpha \geq 0$, C obeys

$$d(x, \phi(x)) \leq \alpha . \frac{\sum_{x' \in S_x} d(x', \phi(x'))}{|S_x|} \quad \forall x \in X. \tag{9}$$

Definition 1.10 *(α-FB Fairness)* (Feature based) [54] A clustering C is said to be α-FB fair if for $\alpha \geq 0$, and similarity set S_x, C obeys

$$|x' \in S_x \text{ and } \phi(x) = \phi(x')| \geq \alpha \quad \forall x \in X. \tag{10}$$

In contrast to the earlier notion, α-PP, α-AG, and α-FB notions allow for an explicit similarity set S_x (perhaps determined through distance or number of matching features). These three notions propound the idea that similar points should be clustered similarly.

Next, we define the Avg-dist notion, which uses well-known clustering stability ideas [77] and the game-theoretic concept of average attraction properties [13]. It induces the individual fairness notion that point x should be closer to its own cluster members than points from other clusters.

Definition 1.11 (*Avg-dist Notion*) [56] A clustering C is said to obey Avg-dist Notion if $\forall x \in C_j$,

$$\frac{1}{|C_j| - 1} \sum_{y \in C_j / x} d(x, y) \leq \frac{1}{|C_i|} \sum_{y \in C_i} d(x, y) \ \forall i \neq j \in [k]. \tag{11}$$

5 Relationships Between Fairness Levels and Their Notions

This section studies the relationship between the different group and individual fairness notions separately, followed by the relationship between the group and individual fairness levels.

5.1 Relationship Between Group Fairness Notions

We empirically examine the relationship between different group fairness notions on two popularly used benchmarking datasets—*adult* and *bank* datasets. The *adult* dataset (census data) contains 21790 males and 10771 females (i.e., *dataset ratio* of 0.49). We use *age*, *fnlwgt*, *education_num*, *capital_gain*, and *hours_per_week* for clustering. The *Bank* dataset (Portuguese Bank records) contains a ternary valued protected attribute—marital status taking values married (24928), single (11568), divorced (4612) forming a *dataset ratio* of 0.18. We use *age*, *duration*, *campaign*, *cons.price.idx*, *euribor3m*, and *nr.employed* for clustering. We fix $k = 10$ and consider k-means clustering.

Many existing algorithms achieving group fairness are either limited to binary-protected groups [11, 29], require extensive hyper-parameter tuning [2, 85], or have high computational complexities [3, 17, 20, 42, 61]. So, we use the polynomial-time algorithm FRAC_{OE} [41], which supports multi-valued protected attribute for the study. FRAC_{OE} is a post-processing technique that solves the fair clustering problem via the fair assignment problem and obtains τ-FAIR fairness. FRAC_{OE} allocates the points from each protected attribute in a round-robin fashion to the centers obtained from the vanilla clustering algorithm. The FRAC_{OE} takes an input value τ_a for each protected group value $a \in S$ (see Definition 1.5). We now ask the question—*Does satisfying the* τ-ratio *fairness notion help to achieve other group fairness notions?* To answer this, we vary τ_a from 0 to $1/k$ (maximum achievable value) and study the induced levels of other group fairness notions. For simplicity, we fix τ_a to be a constant for all possible group values. The results on both datasets are averaged on five independent runs and plotted in Fig. 5 along with standard deviation. From the plots, it is clear that the clusters satisfying τ-ratio fairness also satisfy high BALANCE

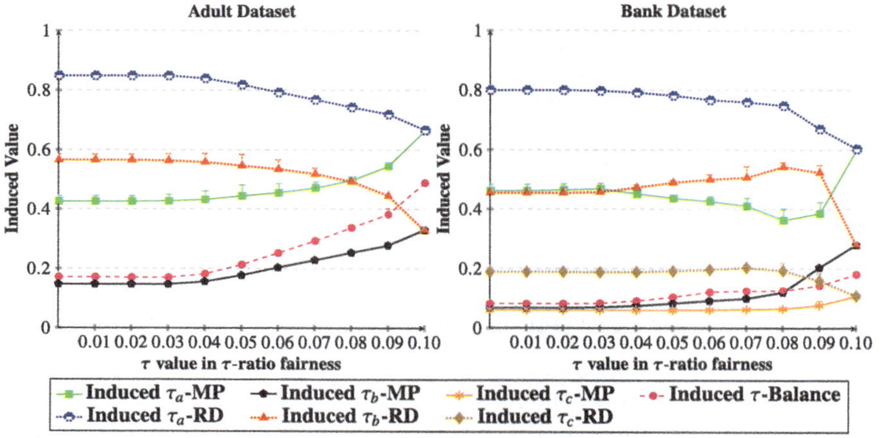

Fig. 5 Induced group fairness values on $k(=10)$-means. (Best viewed in color)

guarantees, maintain lesser restricted dominance, and promote minority protection. The highest value of τ_a leads to maximally balanced clusters.

We execute the linear program (LP) [17] formulated to satisfy MP and RD and observe the τ-ratio fairness level. A remarkable observation is that satisfying MP and RD can lead to a degenerate value of τ-ratio fairness. This will happen when one cluster has very few points from each group (maybe 1), and other clusters contain more points, resulting in highly skewed clusters. Skewed clusters can be problematic in some cases. For example, in problems like direct marketing campaign [16], group fair clustering can be used to segment customers. Highly skewed clusters might not be profitable to invest in for customized solutions. However, using τ-ratio fairness guarantees a minimum number of data points (customers) from each group, i.e., a minimum cluster size while maintaining Balance (induced). This shows that τ-ratio is a stronger notion.

5.2 *Relationship Between Individual Fair Notions*

We next show the connection between individual fairness notions. The results directly follow from definitions.

Result *(1)* If $x \in X$ is α-PP then x is also α-AG.

Proof Since the average value of a set is always larger than the minimum set value, we have:

$$\min_{x' \in S_x} d(x', \phi(x')) \leq \frac{\sum_{x' \in S_x} d(x', \phi(x'))}{|S_x|}.$$

Therefore, if x is α-PP fair, then x is also α-AG fair. □

5.3 Relationship Between Group and Individual Fairness

The two fairness levels arose independently in fair clustering literature. Nonetheless, many real-world applications demand satisfying both group and individual fairness. In direct marketing, the corporate house's diversity policy necessitates group fairness. But at the same time, customers might feel discontented if people in their similarity set belong to a different cluster than their own (hence offering different benefits). Thus, there is a need to study the relationship between the two levels.

Recent attempts [6, 32] explore this direction and propose instances that show the conflicting nature of both the fairness levels, i.e., satisfying one might adversely affect the other. To understand this, consider a dataset with points split across two far-apart clusters, with each cluster containing points from one protected group (as illustrated in Fig. 6a). Group fair clustering will try to place the cluster centers in between the two clusters. On the contrary, the original cluster centers will also serve as optimal individual fair centers when the individual fairness notions depend on distance-based similarity. Thus, showing both fairness as conflicting problems.

We experimentally study the induced individual fairness effect by trying to satisfy group fairness. The reverse trend follows without loss of generality. We use k-means version of FRAC_{OE} algorithm for $k = 10$. We report the maximum deviation value (i.e., α in α-FR) and the fraction of data points satisfying the α-FR with $\alpha = 1$ to the total number of data points. Figure 6b shows the mean and standard deviation over five runs. The plots show that both fairness levels are not strictly in conflict when evaluated on real-world datasets. They both induce certain levels of fairness in the clusters. For both datasets, the number of points having strict individual fairness of

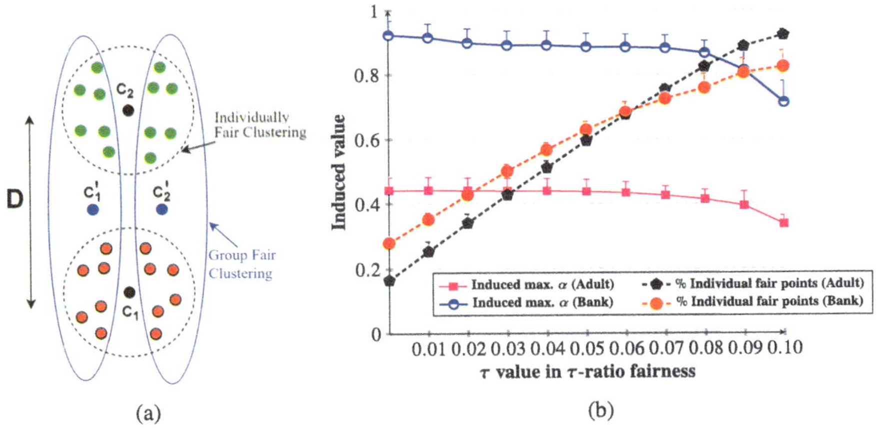

Fig. 6 **a** Example illustrating conflicting group and individual fair clustering. Here C_1, C_2 are individual fair centers separated by large distance D, and C_1', C_2' are group fair centers. **b** Induced α-FR individual fairness values on k(=10)-means

having a center within a given radius increases significantly with an increase in τ. Further, this shows that very few points have large violations, i.e., the maximum α value reported limits to a small set of points.

6 Algorithms and Theoretical Guarantees

6.1 Group Fairness

We now discuss the main ideas of the algorithms and position them with respect to the major theoretical results presented in these works. We also discuss various advantages and disadvantages of each of the approaches.

The foundational work of [29] partitions the data points into small clusters, namely *fairlets*. The paper shows that finding optimal fairlet decomposition is NP-Hard. To find approximate fairlet decomposition, authors use the strategy of solving bipartite matching [36] for maximally balanced clustering and minimum cost flow instances otherwise [9]. The fairlet formed are then merged into k clusters by applying standard (or vanilla) clustering (k-center/k-median) on fairlet centers. The algorithms achieve 4-approx guarantee on τ-BALANCE for k-center and a $(\frac{1}{\tau} + 1 + \sqrt{3} + \epsilon)$-approx guarantee for the k-median objective; where, ϵ is a positive constant. The following are three major shortcomings of this approach: (1) It works only for binary-valued protected groups, (2) It can only achieve the BALANCE same as data points X, and (3) It is not scalable for large datasets. To make the approach scalable, [11] proposed a near-linear time algorithm to compute fairlets using QuadTree data structure [74]. The algorithm computes an embedding with k-median cost approximation of O($hlog(n)$). However, this work is limited to binary-valued protected attributes with k-median clustering objective and can only achieve *dataset ratio*.

Extension to multi-valued protected groups is considered in [20], which proposes a minimum cost-perfect matching (MCPM) algorithm. They provide algorithms for k-center and k-median clustering objectives with 3-approx and $(\rho + 2)$-approx fairness guarantee on τ-BALANCE. However, the algorithm works only when the number of points from each protected group is equal in the dataset X. A similar 14-approx approach using MCPM is proposed in [73] for τ-BALANCE. They further propose a 4-approx method for τ-MP fairness using a reduction to the maximum flow problem. The work is limited to the k-center model. A more general approach with multiple valued protected attributes and with an arbitrary Balance is proposed in [41] by satisfying $\boldsymbol{\tau}$-FAIR. The idea is to convert fair clustering into a fair assignment problem. Motivated by envy freeness (up to one good) literature [5], authors claim that if one knows optimal clustering centers, then the distribution of points in round-robin fashion guarantees that each cluster will be $\boldsymbol{\tau}$-FAIR. This algorithm achieves $2(\rho + 2)$-approx cost guarantees with running complexity of $O(kn \log(n))$.

Among the LP-based techniques, Bera et al. [17] formulate fair clustering as a linear program with $\boldsymbol{\tau}$-RD and $\boldsymbol{\tau}$-MP as constraints. This paper guarantees a maximum fairness violation of at most 3 while simultaneously satisfying ($\rho + 2$)-approx guarantee on the objective cost. Harb et al. [42] extend these guarantees for fair k-center for multi-valued protected groups by formulating LP by restricting the search space for better time complexity. The work by Ahmadian et al. [3] solves fair clustering via $\boldsymbol{\tau}$-RD along with an additional constraint on representative fairness [71]. The authors prove that it is NP-hard to obtain an algorithm better than 2-approx for $\boldsymbol{\tau}$-RD $\in (0, 0.5]$. The proposed algorithm with maximum $O(n^2)$ constraints and variables is 3-approx while the case ($\boldsymbol{\tau}$-RD $= 0.5$) with $O(nk)$ constraints and variables is 12-approx in the clustering objective. The work by Bercea et al. [18] also proposes an LP formulation for $\boldsymbol{\tau}$-MP and $\boldsymbol{\tau}$-RD fairness. The approach achieves a 3, 4.675, and 62.856-approx for k-center, k-median, and k-means, respectively.

Several other works do not provide any theoretical guarantee on the quality of fair clusters. The work by Davidson et al. [32] propose a post-processing technique to impose fairness (in terms of $\boldsymbol{\tau}$-RD and $\boldsymbol{\tau}$-MP) after cluster formation by assigning points from protected groups equally among all k clusters. The goal is to have fewer disagreements between solutions. This approach uses integer linear program (ILP) formulation and uses *total unimodularity* structure of the problem to bypass computational intractability. The complementary problem of minimizing unfairness with maximum allowable clustering cost is considered in Esmaeili et al. [34] using LP formulation. To bound the number of LP iterations, the algorithm exhaustively searches for the feasibility of LPs and chooses the solution with minimum fairness constraints. The integral solution is then constructed using a network flow [18].

Among the regularized based techniques, Ziko et al. [85] propose a *variational* framework where clustering and fairness objectives are simultaneously solved as an optimization problem. This paper uses $\boldsymbol{\tau}$-FE as the fairness notion and breaks the composite problem into convex and concave parts, which are bounded by auxiliary functions. These functions help compute the soft assignment update in the subsequent iteration of k-means, k-median, and N-cut. The authors show that the variational framework has monotonicity and convergence guarantees as Expectation-Maximization (EM) algorithms. The main issue with the approach is the use of data-dependent hyper-parameters. Another important limitation of this approach is that clustering cost deteriorates significantly with an increase in the number of clusters [41]. Liu et al. [61] formulate the problem of fair clustering as a bi-objective optimization problem with $\boldsymbol{\tau}$-BALANCE notion of fairness and prove a sublinear convergence rate. The resulting objective function is non-convex; hence, the solution obtained by stochastic gradient descent does not satisfy any theoretical guarantees on the quality of obtained clusters. Table 1 summarizes all the results.

6.2 Individual Fairness

We now discuss the algorithmic framework and theoretical guarantees of individually fair clustering algorithms. To satisfy α-FR fairness guarantee, a set of *critical balls* is determined [62, 70, 82]. Each critical ball contains a set of points and a *critical* center

Table 1 Categorization of group fairness clustering algorithms. The variable $|I| \leq n$ in [42]. (*source code is available and well tested by us)

	Fairness notions	Time complexity			Cost approximation factor		
		k-means	k-median	k-center	k-means	k-median	k-center
[2]	FairKM	$O(n^2hk)$	×		×		
[3]	$\tau - \text{RD}$	×		Max. variables, constraints: n^2 & nk for $\tau = 0.5$	×		3 & 12 for $\tau = 0.5$
[11]*	$\tau - \text{BALANCE}$	×	$O(hn\log(n))$	×	×	$O(hlog(n))$	×
[17]*	$\tau - \text{BALANCE}$ $\tau - \text{RD}$ & $\tau - \text{MP}$	Max. variables & constraints: $O(n^2)$			$(\rho + 2)$		
[18]	$\tau - \text{RD}$ & $\tau - \text{MP}$	×			3	4.675	62.856
[20]	$\tau - \text{BALANCE}$	$O(n^3T)$	$O(nh)$	$O(nhk)$	$(\rho + 2)$		3
[29]*	$\tau - \text{BALANCE}$	×	$O(T + n^2)$		×	$(\tau + 1 + \sqrt{3} + \epsilon)$	4
[32]	$\tau - \text{RD}$ & $\tau - \text{MP}$	nk regular variables , $2k$ slack variables & $2k + n$ constraints			×		
[34]	$\tau - \text{RD}$ & $\tau - \text{MP}$	Variables & constraints: $O(n^2)$			×		
[41]*	$\tau - \text{BALANCE}$ $\tau - \text{FE}$	$O(kn\log n)$		×	$2(\rho + 2)$		×
[42]*	$\tau - \text{RD}$ & $\tau - \text{MP}$	Max. variables: $\min(2^{k-1}k\mid I\mid, nk)$ & Max.constraints: $k\mid S\mid + \min(2^k\mid I\mid, nk)$			×		
[61]*	$\tau - \text{BALANCE}$	×			×		
[73]	$\tau - \text{BALANCE}$&MP	×		Polynomial	×		14 for τ-BALANCE, 4 for τ-MP
[85]*	$\tau - \text{BALANCE}$ $\tau - \text{FE}$	$O(n^2k^2h)$		×	×		

with the property that each point in a critical ball has a distance less than a pre-defined value from the critical center. In [62], the critical balls are defined such that all the points have distance within $6\alpha r$ to the critical center; here r is defined as the minimum radius containing n/k points from any point. These critical balls are identified using the modified version of greedy approaches proposed in [25, 26]. Next, they use a local search algorithm to improve clustering cost and achieve a bicriteria approximation guarantee[2] of (84, 7)-approx for α-FR and $(O(p), 7)$-approx for general p-norm with k-median as clustering objective.

On similar lines, [82] consider critical balls of radius $2\alpha r$. For fair k-median, authors use k-median algorithm by [80], and for k-center a reduction to standard k-center problem is presented that achieves $(8+\epsilon, 3)$-approx solution where $\epsilon > 0$. For general $p > 1$, a $(16^p, 3)$-approx reduction to matroid facility location problem solved using LP relaxation is proposed. The approximation guarantee is further improved to $(8, 2^{(1+2/p)})$-approx by Negahbani et al. [70], who proposed a fair rounding technique to the optimal LP solution computed using critical centers with radius $2r$.

A feasible solution is not guaranteed for α-PP and α-AG with $\alpha < 2$ [24]. However, any instance with $\alpha \geq 2$ always admits a feasible solution. Even with $\alpha \geq 2$, authors provide an instance where the *price of fairness*[3] without any additional constraint can be arbitrarily bad. For finding feasible centers and fair assignments, the authors provide an algorithm having 5-approx on the fairness guarantee.

Finding α-FB fair clustering is NP-complete even for $k = 2$ [54]. The authors provide a $(1+\epsilon)\text{OPT}(\rho+2)$-approx randomized algorithm solved with the help of LP-relaxation for fair assignment where OPT is optimal fair assignment cost. Similar to α-FB, finding a clustering satisfying *Avg-dist* fairness notion is proved to be NP-Hard even for $X \subseteq \mathbb{R}^2$ ([56]). The authors in [56] further presents a dynamic programming-based solution 1-dimensional setting to find contiguous clusters of target sizes. Table 2 summarizes all the results for individually fair algorithms.

6.3 Extension to Multiple Protected Attributes

Group fairness constraints are also studied under *multiple* multi-valued protected groups setting. For example, an individual can be a female (gender) and native-American (ethnicity). In clustering, this overlap between multiple protected groups is denoted by Δ(=2 in the above example). Both $\boldsymbol{\tau}$-RD and $\boldsymbol{\tau}$-MP can be extended to multiple protected attributes. The work by [17] is also applicable to multiple protected groups with maximum additive violation of $4\Delta + 3$ for $\Delta \geq 2$ (+3 for $\Delta = 1$). The work by [42] provides a similar guarantee.

A notion similar to $\boldsymbol{\tau}$-FE is proposed by [2] for multiple protected attributes. The authors propose a fair k-means algorithm (FairKM) for solving a combined objective

---

[2] (p, q)-approx bicriteria denotes cost approximation of p and fairness approximation of q.

[3] Ratio of clustering objective value under fairness constraint to the standard objective value.

Table 2 Categorization of individual fair clustering algorithms. $\mathcal{OPT}$ is optimal for fair assignment cost in [54] and $\epsilon > 0$. (*source code is available and well tested by us)

	Fairness notion	Time complexity	Cost approximation factor		
			k-means	k-median	k-center
[24]*	α-PP	$\times$	5-approx w.r.t fairness		
	α-AG				
[54]	α-FB	Polynomial	$((1+\epsilon)\text{OPT}(\rho+2))$		
[56]	Avg. dist based	$O(n^3k)$	$\times$		
[62]*	α-FR	$O(k^5n^4)$	$(O(p), 7)$	(84, 7)	$(O(p), 7)$
[70]*	α-FR	$O(k.n^4)$	$(8, 2^{1+\frac{2}{p}})$		
[82]	α-FR	Polynomial	$(16^p, 3)$	$(8+\epsilon, 3)$	$(8+\epsilon, 3)$

function of minimizing objective cost along with a deviation in this modified notion of fairness. The algorithm, however, is sensitive to the trade-off parameter, needs extensive tuning, and requires minimization of a non-convex function.

6.4 Fair Algorithms Under Different Setting

In a slightly different line of work, [28] proposes a pre-processing technique that augments the data with a small number of points called *antidotes*, encoding the fairness requirement. It provides necessary restrictions to the clustering algorithm, ultimately leading to fair clustering. When there is imperfect knowledge of protected group membership [35, 67] provide a probabilistic clustering framework. The central idea is that it assumes a probability distribution over the possible value of the protected attribute. In a slightly different context of streaming algorithms, group fairness is studied by [14, 46, 75, 76] using *coresets*. Finally, [58] studies group fairness under capacitated settings.

6.5 Deep Fair Clustering

Recent efforts are being carried out to further improvise clustering efficiency with deep methods. However, these methods can further degrade fairness in a trade-off for objective cost. To address this issue, [83] proposes group-level fairness for a multi-valued protected attribute. Initially, the idea is to generate a probabilistic assignment using a vanilla clustering neural network. To incorporate τ-BALANCE, they formulate an integer LP that tries to modify cluster assignments minimally. Finally, vanilla and fair assignment are exploited to train a fair clustering network using contrastive learning. The work in [84] proposes a deep fair clustering embedding model that jointly trains objective cost and τ-BALANCE. For achieving equal representation in each cluster, centroids are made equidistant to fairoid (fairness centroid).[4]

7 Discussion and Open Problems

In this chapter, we surveyed results in fair clustering literature focusing on two fundamental levels of fairness; group and individual fairness. This chapter reveals that many proposed notions are interconnected and often imply each other. We identified the relationships between mathematical formulations of these fairness notions. We also provided a categorization of fair clustering algorithms across multiple dimensions, such as implementation stage, solution approaches, and time complexity. We also discussed different fair clustering algorithms, surveyed their performance guarantees, and identified their limitations. We now provide research gaps seen in Tables 1, 2, and several conceptual future directions in fair clustering.

Group + Individual fairness—Historically, these two fairness levels have been studied separately. However, many real-world problems require fulfilling both these levels together, an aspect that needs immediate attention from the research community. Though some recent studies (namely [6, 32]) explore this direction, a large-scale effort is required to better understand relationships between different mathematical formulations, design efficient algorithms with provable group+individual fairness guarantees, and demonstrate limitations of each combination of fairness notions.

Pareto frontier Analysis—The Pareto-optimal frontier provides a complete characterization of the trade-off between multiple objectives in an optimization problem. Many current studies in fair clustering consider fairness requirements as hard constraints or provide guarantees based on data-dependent constraints. A study of the Pareto frontier between fairness and clustering objective cost would help theoreticians and practitioners understand situations in which the trade-off is of most practical significance. The extent to which such a characterization is possible and the study of algorithmic frameworks to achieve this trade-off is an interesting open problem.

Generalizations to multiple attributes—We briefly reviewed generalization of τ-RD and τ-MP fairness notions to multiple overlapping attributes setting in Sect. 6.3. Extending other fairness notions to multiple protected attributes and understanding the relationship between these notions in multiple attribute settings is an important and practically relevant research direction. A further extension to simultaneously meet different group fairness notions is also an important open problem.

Gaming and incentives design—Throughout this chapter and in most of the fair clustering literature, it is assumed that the data generation process is noise-free and non-strategic. Seen as a natural extension of *strategic classification* [43] in a clustering framework, *strategic clustering* (See, [79]) has many practical applications. For

[4] Mean center of all points belonging to a single color (say red points) in the dataset.

instance, in a consumer segmentation application where agents have preferences over segments (i.e., clusters) in which they are assigned, may *game* the algorithm by misreporting their data to obtain the desired assignment. This misreporting may result in a significant loss in the objective function, and consequently, the fairness guarantees may fail to hold. Studying incentives to elicit truthful reports and designing robust gaming fair clustering algorithms is an interesting future work.

Fair clustering in federated setting—In a federated learning system (FLS), several parties train machine learning models collaboratively without exchanging their raw data. The clients send their local updates to the global server, which aggregates them and communicates back the averaged updates to the local clients [64]. Federated Learning leads to smarter models (as they are trained on large data) while ensuring privacy. There are a few works that consider fair classification under federated settings [53], but fair clustering is still limited to the centralized environment. A possible extension could be to design privacy-preserving fair clustering algorithms, which can potentially be used in recommendation systems or data summarization.

Group fair algorithms under highly skewed distributions—In many real-world applications, disadvantaged groups are present in quite a small fraction compared to advantaged groups. For example, trans-gender against males and minority religions in contrast to widely adopted religions in a given geographical region and so on. In a supervised learning setting, it has been observed that fairness-aware constraints that tend to equalize the performance across the groups may lead to non-uniform degradation in performance for highly skewed datasets [12]. Current state-of-art algorithms analyze mostly synthetic or real-world datasets with a good distribution of all protected groups. Thus, investigating the performance of existing algorithms in datasets exhibiting highly skewed distributions of protected group values is another open research problem.

References

1. Abbasi M, Bhaskara A, Venkatasubramanian S (2021) Fair clustering via equitable group representations. In: ACM FAccT, pp 504–514. https://doi.org/10.1145/3442188.3445913
2. Abraham SS, Padmanabhan D, Sundaram SS (2020) Fairness in clustering with multiple sensitive attributes. In: EDBT/ICDT joint conference, pp 287–298
3. Ahmadian S, Epasto A, Kumar R, Mahdian M (2019) Clustering without over-representation. In: SIGKDD, pp 267–275. https://doi.org/10.1145/3292500.3330987
4. Ahmadian S, Epasto A, Kumar R, Mahdian M (2020) Fair correlation clustering. In: International conference on artificial intelligence and statistics. PMLR, pp 4195–4205
5. Amanatidis G, Aziz H, Birmpas G, Filos-Ratsikas A, Li B, Moulin H, Voudouris AA, Wu X (2022) Fair division of indivisible goods: a survey. arXiv:2208.08782
6. Anderson N, Bera SK, Das S, Liu Y (2020) Distributional individual fairness in clustering. arXiv:2006.12589
7. Anegg G, Angelidakis H, Kurpisz A, Zenklusen R (2020) A technique for obtaining true approximations for k-center with covering constraints. In: International conference on integer programming and combinatorial optimization. Springer, pp 52–65
8. Anegg G, Koch LV, Zenklusen R (2022) Techniques for generalized colorful k-center problems. arXiv:2207.02609

9. Asano T, Asano Y (2000) Recent developments in maximum flow algorithms. J Oper Res Soc Jpn 43(1):2–31
10. Bacelar M (2021) Monitoring bias and fairness in machine learning models: a review. ScienceOpen Preprints
11. Backurs A, Indyk P, Onak K, Schieber B, Vakilian A, Wagner T (2019) Scalable fair clustering. In: ICML, pp 405–413
12. Balashankar A, Lees A, Welty C, Subramanian L (2019) What is fair? exploring pareto-efficiency for fairness constrained classifiers. arXiv:1910.14120
13. Balcan MF, Blum A, Vempala S (2008) A discriminative framework for clustering via similarity functions. In: ACM STOC, pp 671–680
14. Bandyapadhyay S, Fomin FV, Simonov K (2020) On coresets for fair clustering in metric and euclidean spaces and their applications. arXiv:2007.10137
15. Bandyapadhyay S, Inamdar T, Pai S, Varadarajan K (2019) A constant approximation for colorful k-center. arXiv:1907.08906
16. Banerjee A, Ghosh J (2006) Scalable clustering algorithms with balancing constraints. Data Min Knowl Discov 13(3):365–395
17. Bera S, Chakrabarty D, Flores N, Negahbani M (2019) Fair algorithms for clustering. In: NeurIPS, pp 4954–4965
18. Bercea IO, Groß M, Khuller S, Kumar A, Rösner C, Schmidt DR, Schmidt M (2018) On the cost of essentially fair clusterings. arXiv:1811.10319
19. Biddle D (2017) Adverse impact and test validation: a practitioner's guide to valid and defensible employment testing. Routledge
20. Böhm M, Fazzone A, Leonardi S, Schwiegelshohn C (2020) Fair clustering with multiple colors. arXiv:2002.07892
21. Brubach B, Chakrabarti D, Dickerson J, Khuller S, Srinivasan A, Tsepenekas L (2020) A pairwise fair and community-preserving approach to k-center clustering. In: ICML, pp 1178–1189
22. Brubach B, Chakrabarti D, Dickerson JP, Srinivasan A, Tsepenekas L (2021) Fairness, semi-supervised learning, and more: a general framework for clustering with stochastic pairwise constraints. In: Proceedings of the AAAI conference on artificial intelligence, vol 35, pp 6822–6830
23. Byrka J, Pensyl T, Rybicki B, Srinivasan A, Trinh K (2014) An improved approximation for k-median, and positive correlation in budgeted optimization. In: ACM-SIAM SODA, pp 737–756
24. Chakrabarti D, Dickerson JP, Esmaeili SA, Srinivasan A, Tsepenekas L (2021) A new notion of individually fair clustering: α-equitable k-center. arXiv:2106.05423
25. Chan THH, Dinitz M, Gupta A (2006) Spanners with slack. In: European symposium on algorithms. Springer, pp 196–207
26. Charikar M, Makarychev K, Makarychev Y (2010) Local global tradeoffs in metric embeddings. SIAM J Comput 39(6):2487–2512
27. Chhabra A, Masalkovaitė K, Mohapatra P (2021) An overview of fairness in clustering. IEEE Access
28. Chhabra A, Singla A, Mohapatra P (2021) Fair clustering using antidote data. arXiv:2106.00600
29. Chierichetti F, Kumar R, Lattanzi S, Vassilvitskii, S.: Fair clustering through fairlets. In: NeurIPS, pp. 5036–5044 (2017)
30. Chlamtáč E, Makarychev Y, Vakilian A (2022) Approximating fair clustering with cascaded norm objectives. In: Proceedings of the 2022 annual ACM-SIAM symposium on discrete algorithms (SODA). SIAM, pp 2664–2683
31. Dastin J (2018) Amazon scraps secret ai recruiting tool that showed bias against women. https://www.reuters.com/article/us-amazon-com-jobs-automation-insight-idUSKCN1MK08G. Accessed 15-August-2021
32. Davidson I, Ravi S (2020) Making existing clusterings fairer: algorithms, complexity results and insights. AAAI 34(04):3733–3740. https://doi.org/10.1609/aaai.v34i04.5783. ojs.aaai.org/index.php/AAAI/article/view/5783

33. Dwork C, Hardt M, Pitassi T, Reingold O, Zemel R (2012) Fairness through awareness. In: ITCS, pp 214–226
34. Esmaeili S, Brubach B, Srinivasan A, Dickerson J (2021) Fair clustering under a bounded cost. In: NeurIPS
35. Esmaeili S, Brubach B, Tsepenekas L, Dickerson J (2020) Probabilistic fair clustering. In: NeurIPS, pp 12743–12755
36. Galil Z (1986) Efficient algorithms for finding maximum matching in graphs. ACM Comput Surv (CSUR) 18(1):23–38
37. Ghadiri M, Samadi S, Vempala S (2021) Socially fair k-means clustering. In: ACM FAccT, pp 438–448
38. Ghassami A, Khodadadian S, Kiyavash N (2018) Fairness in supervised learning: an information theoretic approach. In: IEEE ISIT, pp 176–180
39. Gonzalez TF (1985) Clustering to minimize the maximum intercluster distance. Theor Comput Sci 38:293–306
40. Goyal D, Jaiswal R (2021) Tight fpt approximation for socially fair clustering. arXiv:2106.06755
41. Gupta S, Ghalme G, Krishnan NC, Jain S (2021) Efficient algorithms for fair clustering with a new fairness notion. arXiv:2109.00708
42. Harb E, Lam HS (2020) Kfc: a scalable approximation algorithm for k- center fair clustering. In: NEURIPS, pp 14509–14519
43. Hardt M, Megiddo N, Papadimitriou C, Wootters M (2016) Strategic classification. In: ITCS, ITCS '16. Association for Computing Machinery, New York, NY, USA, pp 111–122
44. Hochbaum DS, Shmoys DB (1986) A unified approach to approximation algorithms for bottleneck problems. J ACM (JACM) 33(3):533–550
45. Hong W, Zheng S, Wang H (2013) A job recommender system based on user clustering. J Comput 8(8) (2013)
46. Huang L, Jiang S, Vishnoi N (2019) Coresets for clustering with fairness constraints. In: NeurIPS, pp 7589–7600
47. Jain AK, Murty MN, Flynn PJ (1999) Data clustering: a review. ACM Comput Surv 31(3):264–323. https://doi.org/10.1145/331499.331504. doi.org/10.1145/331499.331504
48. Jia X, Sheth K, Svensson O (2020) Fair colorful k-center clustering. In: International conference on integer programming and combinatorial optimization. Springer, pp 209–222
49. Jones M, Nguyen H, Nguyen T (2020) Fair k-centers via maximum matching. In: ICML, pp 4940–4949
50. Julia A, Larson J (2016) Propublica machine bias. https://www.propublica.org/article/machine-bias-risk-assessments-in-criminal-sentencing. Accessed 13-August-2021
51. Julia A, Larson J, Mattu S, Kirchner L (2016) Propublica–machine bias. https://www.propublica.org/article/machine-bias-risk-assessments-in-criminal-sentencing. Accessed 13-August-2021
52. Jung C, Kannan S, Lutz N (2020) Service in your neighborhood: fairness in center location. Foundations of responsible computing
53. Kanaparthy S, Padala M, Damle S, Gujar S (2022) Fair federated learning for heterogeneous data. In: Joint CODS-COMAD, pp 298–299. https://doi.org/10.1145/3493700.3493750
54. Kar D, Medya S, Mandal D, Silva A, Dey P, Sanyal S (2021) Feature-based individual fairness in k-clustering. arXiv:2109.04554
55. Kleindessner M, Awasthi P, Morgenstern J (2019) Fair k-center clustering for data summarization. In: ICML, pp 3448–3457
56. Kleindessner M, Awasthi P, Morgenstern J (2020) A notion of individual fairness for clustering. arXiv:2006.04960
57. Kurdija AS, Afric P, Sikic L, Plejic B, Silic M, Delac G, Vladimir K, Srbljic S (200) Candidate classification and skill recommendation in a cv recommender system. In: International conference on AI and mobile services. Springer, pp 30–44
58. Le Quy T, Roy A, Friege G, Ntoutsi E (2021) Fair-capacitated clustering. In: EDM, pp 407–414

59. Li B, Li L, Sun A, Wang C, Wang Y (2021) Approximate group fairness for clustering. In: ICML, pp 6381–6391. http://proceedings.mlr.press/v139/li21j.html
60. Li S, Svensson O (2016) Approximating k-median via pseudo-approximation. SIAM J Comput 45(2):530–547
61. Liu S, Vicente LN (2021) A stochastic alternating balance k-means algorithm for fair clustering. arXiv:2105.14172
62. Mahabadi S, Vakilian A (2020) Individual fairness for k-clustering. In: ICML, pp 6586–6596
63. Makarychev Y, Vakilian A (2021) Approximation algorithms for socially fair clustering. In: Belkin M, Kpotufe S (eds) COLT. https://proceedings.mlr.press/v134/makarychev21a.html
64. McMahan HB et al (2021) Advances and open problems in federated learning. Found Trends® Mach Learn 14(1)
65. Mehrabi N, Morstatter F, Saxena N, Lerman K, Galstyan A (2021) A survey on bias and fairness in machine learning. ACM Comput Surv 54(6). https://doi.org/10.1145/3457607
66. Mhasawade V, Zhao Y, Chunara R (2021) Machine learning and algorithmic fairness in public and population health. Nat Mach Intell 3(8):659–666
67. Micha E, Shah N (2020) Proportionally fair clustering revisited. In: ICALP
68. Moulin H (2004) Fair division and collective welfare. MIT Press
69. Nedlund E (2019) Apple card is accused of gender bias.here's how that can happen. https://edition.cnn.com/2019/11/12/business/apple-card-gender-bias/index.html. Accessed 1-November-2022
70. Negahbani M, Chakrabarty D (2021) Better algorithms for individually fair k-clustering. In: NeurIPS
71. Padmanabhan D, Abraham SS (2020) Representativity fairness in clustering. In: 12th ACM conference on web science. http://dx.doi.org/10.1145/3394231.3397910
72. Padmanabhan D (2020) Whither fair clustering? In: AI for social good: CRCS workshop
73. Rösner C, Schmidt M (2018) Privacy preserving clustering with constraints. In: 45th international colloquium on automata, languages, and programming (ICALP 2018). Schloss Dagstuhl-Leibniz-Zentrum fuer Informatik
74. Samet H (1984) The quadtree and related hierarchical data structures. ACM Comput Surv (CSUR) 16(2):187–260
75. Schmidt M, Schwiegelshohn C, Sohler C (2019) Fair coresets and streaming algorithms for fair k-means. In: International workshop on approximation and online algorithms. Springer, pp 232–251
76. Schmidt M, Wargalla J (2021) Coresets for constrained k-median and k-means clustering in low dimensional Euclidean space. arXiv:2106.07319
77. Sharifi-Malvajerdi S, Kearns M, Roth A (2019) Average individual fairness: algorithms, generalization and experiments. In: NeurIPS, pp 8242–8251
78. Song M, Rajasekaran S (2010) Fast algorithms for constant approximation k-means clustering. Trans Mach Learn Data Min 3(2):67–79
79. Stoica AA, Papadimitriou C (2018) Strategic clustering. http://www.columbia.edu/as5001/strategicclustering.pdf. Accessed 22-January-2022
80. Swamy C (2016) Improved approximation algorithms for matroid and knapsack median problems and applications. ACM Trans Algorithms 12(4). https://doi.org/10.1145/2963170
81. Thejaswi S, Ordozgoiti B, Gionis A (2021) Diversity-aware k-median: clustering with fair center representation. In: Joint European conference on machine learning and knowledge discovery in databases. Springer, pp 765–780
82. Vakilian A, Yalçıner M (2021) Improved approximation algorithms for individually fair clustering. arXiv:2106.14043
83. Wang B, Davidson I (2019) Towards fair deep clustering with multi-state protected variables. arXiv:1901.10053
84. Zhang H, Davidson I (2021) Deep fair discriminative clustering. arXiv:2105.14146
85. Ziko IM, Yuan J, Granger E, Ayed IB (2021) Variational fair clustering. In: AAAI, pp 11202–11209

Temporal Fairness in Online Decision-Making

Swati Gupta, Vijay Kamble, and Jad Salem

Abstract In many real-world decision-making problems, inputs and data evolve over time, and decisions must be made in this dynamic (online) framework. For example, one might wish to screen resumes in real-time as they are submitted; in this setting, the feedback used to make the screening decisions (such as ultimate hiring decisions on past resumes) might be provided over time as well. While fairness concerns have received significant attention in offline settings, relatively little is known about fairness in these dynamic settings; this begs the question, what does it mean to be fair in online decision-making? This question is complicated because stringent constraints can prevent good decisions later on due to a few bad decisions early on, and further, the data which is used to justify decisions varies over time. Given these challenges, there is a need for new ideas to understand the trade-offs between fairness and utility in dynamic decision-making settings. This chapter surveys existing notions of temporal fairness and reviews recent work in this growing area.

1 Introduction

With the increased adoption of automated decision-making practices in domains ranging from recommendation systems [3] and retail pricing [2] to banking [12], criminal justice [18], and healthcare [16], there has been a corresponding surge of interest in the academic community and among policymakers alike on the topic of *algorithmic fairness*, i.e., ensuring that humans are treated fairly and equitably by

S. Gupta
Massachusetts Institute of Technology, 77 Massachusetts Avenue, Cambridge, MA, USA
e-mail: swatig@mit.edu

V. Kamble
University of Illinois at Chicago, 601 S Morgan St, 60607 Chicago, IL, USA
e-mail: kamble@uic.edu

J. Salem (✉)
U.S. Naval Academy, 121 Blake Road, Annapolis, MD, USA
e-mail: jsalem@usna.edu

A. Mukherjee et al. (eds.), *Ethics in Artificial Intelligence: Bias, Fairness and Beyond*,
Studies in Computational Intelligence 1123,
https://doi.org/10.1007/978-981-99-7184-8_3

these systems [4, 6]. As this discourse develops, various task-specific notions of fairness have been introduced. This is natural since the social backdrop of, say, loan administration is very different from that of job applicant screening [19]. Hence, the notions of fairness appropriate in different decision-making settings such as pricing, applicant screening, loan administration, etc., may be different. For instance, one may wish to equalize false negative rates of a cancer diagnosis algorithm across racial groups to ensure that no racial group is underdiagnosed. For a criminal risk prediction algorithm, on the other hand, it may be desirable to equalize false positive rates across groups to ensure that no group is under disproportionate scrutiny.

These settings, however, commonly involve decision-making *over a period of time*. A large company will receive job applications consistently over time and must make screening decisions as they arrive. Loan granting decisions are similarly made over time as applications arrive. Goods are priced periodically. Not only are decisions made over time in each of these cases, but data changes over time as well. For example, as loan grantees pay back or default on their loans, this information can be used to inform future loan granting decisions. This raises important, complex questions about how existing (static) notions of fairness should be adapted to settings where data and decisions are *dynamic*.

To begin with, the introduction of time leads to **normative questions** about the meaning of fairness. Suppose that in the context of applicant screening, we have convinced ourselves that equalizing selection rates across groups (i.e., demographic parity) is "fair" in the static setting. What, then, should fairness mean in a dynamic setting, where decisions are made over time, and feedback is received over time? Is it fair to equalize selection rates across groups across all time periods? This might involve constraining decisions made today by decisions made 10 years ago. For example, in a profession dominated by men, striving for demographic parity *across all time* would require selecting no men until cumulative selections (or selection rates) are equal. If we instead wanted demographic parity within a sliding window, we could require those selections (or selection rates) over the past, say, year, be equal. Which of these approaches, if any, is a fair way to make applicant screening decisions over time?

To complicate matters further, the introduction of time results in **fairness-learnability** trade-offs. In many sequential decision-making problems, the "rewards" of different decisions are often not known in advance and must be learned over time through exploration. For example, in dynamic pricing with demand learning, the maximum-revenue price point is not known in advance but is estimated over time by experimenting at different prices. If a fairness constraint is very stringent, our ability to learn and converge to an optimal decision can be hindered, and performance can suffer.

Given these complex normative and technical challenges, the problem of choosing fairness constraints in dynamic settings is difficult. In this chapter, we will provide a framework for dynamic adaptations of static fairness constraints and explore these challenges further.

2 Fairness in Static Decision-Making

Consider a static decision-making scenario, where decisions must be given all at once to a set of stakeholders. Each stakeholder has a context $c \in C$, and for the sake of consistency, we require that stakeholders with the same context receive the same decision. For example, if the task is multi-segment pricing, each customer segment (e.g., "youth") could be a context. The decisions lie in the set $\mathcal{X}$ (e.g., the range of prices that can be offered to a customer, or accept/reject decisions for applicant-screening).

There is a principal (such as a firm or a platform) who chooses a decision rule that maps contexts to decisions. We represent such a rule by the function $x : C \to \mathcal{X}$. There is an evaluator function $\mathcal{U}$ that measures the utility from a decision rule, which represents some operational performance measures such as social welfare, efficiency, principal utility, etc.

Example 1

Suppose job applicants take two ability tests (each scored out of 10) as part of their application, and screening decisions are made based on the sum of their scores. In this case, $C = \{0, 1, 2, \ldots, 20\}$, and for each combined score n, a binary screening decision is made: $x(n) \in \{0, 1\} = \mathcal{X}$ (or alternatively, one can assign to each applicant a *probability of selection* in $\mathcal{X} = [0, 1]$). In this case, the decision rule is any element in $\mathcal{X}^{|C|} = \{0, 1\}^{21}$.

Example 2

Loan granting can be modeled similarly to applicant screening as described in Example 2. In this case, a decision of $x \in [0, 1]$ would correspond to a probability of x of granting the loan.

Fairness constraints can take many forms. Some are *group-based*, requiring similar qualities (e.g., false positive rates) across groups, while others are *contextual*, requiring some notion of consistency (e.g., individual fairness) across contexts. Some constraints are strictly *outcomes-based*, typically requiring that different groups receive similar proportions of positive decisions, while others allow for disparate outcomes depending on data (e.g., equalizing false positive rates across groups does not ensure equality of outcomes).

To generally capture these different types of constraints, we need a model that accounts for (1) the current decisions and (2) data, in the form of past contexts, decisions, and feedback (e.g., this data could be the training data on which an ML model is based). To that end, let $\mathcal{D}$ be some space in which the historic data lives, and suppose there are k functions $F_i : \mathcal{X}^{|C|} \times \mathcal{D} \to \mathbb{R}$ for $i = 1, \ldots, k$. These functions map (a) the decision rule and (b) the available historical data to a real number. We then assume that the fairness notion requires that

$$F_i(x, D) \leq 0 \text{ for all } i = 1, \ldots, k.$$

Example 3

Suppose we want to approximately ensure equality of outcomes in applicant screening. In particular, suppose each applicant i has a context c_i in some finite context space, the context space can be partitioned into two groups C_1 and C_2 (e.g., those over 40 years old, and those under 40) where $C_1 \cup C_2 = C$ and $C_1 \cap C_2 = \emptyset$, and we want the selection rates to differ by a factor of at most $4/5$. The data available to us is the sequence D of contexts corresponding to the applicant pool; i.e., $D = (c_1, c_2, \ldots, c_n)$, where n is the number of applicants. Now let A_i be the set of applicants with context in C_i, for $i = 1, 2$, and assume that $A_1, A_2 \neq \emptyset$. In this case, the following constraints would achieve this goal:

$$F_1(x, D) = \frac{\sum_{i \in A_1} \frac{x(c_i)}{|A_1|}}{\sum_{i \in A_2} \frac{x(c_i)}{|A_2|}} - \frac{5}{4} \leq 0$$

$$F_2(x, D) = \frac{\sum_{i \in A_2} \frac{x(c_i)}{|A_2|}}{\sum_{i \in A_1} \frac{x(c_i)}{|A_1|}} - \frac{5}{4} \leq 0.$$

Note that the historic decisions and feedback were not used in these constraints: our only goal was to equalize outcomes in the current decision rule, irrespective of past decisions and feedback.

Example 4

Suppose we want to ensure some notion of comparative fairness in movie ticket pricing. In particular, suppose we have a finite context space $C = \{\text{Y}, \text{AS}, \text{AN}\}$, where Y represents youths, AS represents adult students, and AN represents adult non-students. For each context $c \in C$, we want to choose a price $x(c) \in [0, 10]$. Suppose we want the youth price and the adult student price to be at most the adult non-student price, and we want the youth price and the adult student price to be within \$1 of each other. Then we can require the following constraints:

$$F_1(x, D) = x(\text{Y}) - x(\text{AN}) \leq 0$$

$$F_2(x, D) = x(\text{AS}) - x(\text{AN}) \leq 0$$

$$F_3(x, D) = x(\mathrm{Y}) - x(\mathrm{AS}) - 1 \leq 0$$
$$F_4(x, D) = x(\mathrm{AS}) - x(\mathrm{Y}) - 1 \leq 0$$

The static optimization problem of maximizing the utility of a decision rule subject to fairness constraints can then be expressed as follows:

$$\text{maximize } \mathcal{U}(x) \quad (1)$$
$$\text{subject to } F_i(x, D) \leq 0 \text{ for all } i = 1, \ldots, k. \quad (2)$$

When $\mathcal{U}$ and F_i are well-behaved functions, one can find a solution using standard optimization techniques. For example, if $\mathcal{U}$ and F_i can be expressed as linear functions of a finite number of variables, then the simplex method can be used to find the optimal fair decision rule.

3 Ensuring Fairness in Dynamic Settings

The introduction of time in the above setup makes it more interesting. Suppose that the principal chooses decision rules over a time horizon of multiple periods, $t \in \{1, \ldots, T\} = [T]$. Let's denote the decision rule at time t as $x_t : C \to \mathcal{X}$ and the available data at time t as $D_{1:t}$. First, note that if there is no reason to choose different decision rules across time periods, e.g., in settings where the utility function $\mathcal{U}$ is unchanging and known to the principal, then the time dimension is redundant since the principal can simply choose the same static-optimal fair decision rule at each time. However, there may be reasons that may make the principal want to vary decision rules over time. For example, $\mathcal{U}$ itself could be time-dependent in certain settings. Or $\mathcal{U}$ could be unknown to the principal, and the optimal fair decision rule must be learned over time. Or feedback in $D_{1:t}$ can accrue over time (as t grows), influencing which decisions the principal views as fair. In this section, we will explore potential concerns in defining a temporal fairness constraint.

3.1 Temporal Fairness and Memory

A key question in extending static notions of fairness to dynamic settings is that of *memory*: how far back should one look? Suppose, for example, that we wanted to satisfy comparative fairness in probabilities of selection in applicant screening: similar contexts should receive similar decisions, and the same context should always receive the same decision. If our memory extends all the way to the first set of decisions made, then an applicant with context c today must be treated the same as an applicant with context c was treated on day 1. While this would unambiguously satisfy the goal of comparative fairness, it would arguably be too constraining: one bad decision early on can prevent good decisions later on.[1] We categorize temporal fairness notions based on the amount of memory used, as we discuss next.

No memory: fairness within each period. The first and most basic possibility in this regard is to simply ignore the temporal aspect and ensure that the desired static fairness constraints are satisfied independently in each time period. That is, we require that $F_i(x_t, D_{t:t}) \leq 0$ for all $i = 1, \ldots, k$ and all times $t = 1, \ldots, T$, where $D_{t:t}$ is the data from the current time period only (e.g., contexts of the current batch of applicants, not contexts, decisions, or feedback from previous batches). In some situations, this may suffice from a fairness perspective. For instance, it may be acceptable that the salaries of women fell due to some event (such as a pandemic) if men's salaries proportionally fell as well to ensure that the salaries are always equitable. From a technical perspective, such fairness constraints can often be employed in practice using standard optimization techniques. For example, assuming that the concerned functions are well-behaved (e.g., convex or concave) and the feasible region is structured (e.g., polyhedral), the theory of online convex optimization can readily handle the problem of optimizing the utility over time while ensuring that the decision rules are feasible at each time. However, in many scenarios, such time-independent notions may be insufficient; e.g., it may be difficult to justify that a small business loan application was approved today, but an applicant with an identical profile was rejected the next day. When consistency across time is desired, *full memory* or *partial memory* notions may be appropriate, as discussed next.

Full memory: fairness across all time. The second possibility lies on the opposite end of the spectrum, requiring that the fairness of a decision must be satisfied with respect to decisions across all times. That is for each $i = 1, \ldots, k$,

$$F_i\left(x_t, D_{1:t}\right) \leq 0 \text{ for all } i = 1, \ldots, k. \tag{3}$$

In many cases, full memory constraints are quite stringent (cf. Example 5 and Sect. 3.2), since they can disallow any change in decisions over time [8]. Such a requirement is impractical in scenarios where changes in the decision rule may be practically necessary across time, e.g., to learn an unknown utility function; such changes may even be acceptable from a fairness standpoint, e.g., at least *increases* in salaries over time are generally acceptable (and often necessary in view of inflation).

[1] We chose the term "memory" to allude to the one-sided perspective of the decision-maker in an online setting: the current decision can only reasonably be constrained by *previous* decisions if the future decisions are unknown. One could also think of any current decision x_t as imposing a constraint on the future decisions (e.g., a lower bound or an upper bound, especially in the case of fairness across all time). With this interpretation, the term "memory" loses its motivation. While this latter interpretation may be salient from a theoretical perspective of the feasibility of policies, we find the former interpretation to be more compelling in dynamic (online) settings where future decisions are unknown and constraining these is not natural.

Example 5

Consider an applicant-screening scenario in which we must assign a probability of selection $x \in \mathcal{X} = [0, 1]$ to each context $c \in C \subset \mathbb{R}^N$, and we want our decision rule to be Lipschitz (i.e., to satisfy individual fairness) with respect to some metric d. If we opt for full memory, then decisions made at time t are constrained by decisions made at time 1. If context c was observed at time 1 and a bad decision was made, then decisions made on contexts in a small ball around c will be forced to be suboptimal in all future time periods. Even allowing for slack in the constraints (i.e., replacing the constraint $|x(c_1) - x(c_2)| \leq d(c_1, c_2)$ with $|x(c_1) - x(c_2)| \leq d(c_1, c_2) + \varepsilon$) does not solve this issue, as the initial decision can be off by more than ε.

Partial memory. There are several possibilities between these two extremes. For example, one can impose a sliding window constraint, where the fairness constraint can be based on data from the previous m time periods. Similarly, one can impose a time-decay constraint, where recent decisions are weighted more heavily than older decisions. Importantly, this can help prevent the issues discussed in the previous example.

Example 6

Suppose we are making cancer diagnosis decisions at a local clinic, and we would like to approximately equalize false negative rates across the disjoint groups C_1 and C_2, where $C_1 \cup C_2 = C$. However, the air quality in the city in question has degraded over the past several years, and this change is suspected to impact the relationship between contexts and cancer risk. Due to this changing socio-environmental landscape, we want to constrain decisions only based on the previous m decisions. Suppose that the available data $D_{t-m:t} = (c_{t-m}, \ldots, c_t, x_{t-m}(c_{t-m}), \ldots, x_t(c_t), y_{t-m}, \ldots, y_t)$ contains the previous m contexts, the previous m decisions, and the previous m outcomes (whether or not the individual was eventually diagnosed with cancer). In this case, the sliding window false negative rate for group C_i at time t (i.e., after the tth decision) is

$$\text{FNR}(C_i) = \frac{\sum_{\substack{j=\max\{1,t-m\},\ldots,t \\ c_j \in C_i}} \mathbb{1}[x_j(c_j) = 0, \ y_j = 1]}{|\{\max\{1, t-m\} \leq j \leq t : c_j \in C_i, y_j = 1\}|}.$$

To approximately enforce equal false negative rates, we require that at each time t,

$$F_1(x_t, D_{\max\{1,t-m\}:t}) = \text{FNR}(C_1) - \text{FNR}(C_2) - \varepsilon \leq 0$$
$$F_2(x_t, D_{\max\{1,t-m\}:t}) = \text{FNR}(C_2) - \text{FNR}(C_1) - \varepsilon \leq 0.$$

3.2 Temporal Fairness and Learnability

As alluded to in Sect. 3.1, there can be tensions between fairness and learnability in online learning problems. For example, consider the scenario in Example 5. We can quantify the performance of an online learning algorithm using its *regret*, defined as follows. Suppose that at time t, we observe a context $c_t \in C$ and choose a decision rule $x_t : C \to \mathcal{X}$. Given a utility function $\mathcal{U}$, the regret up to time T is defined as

$$\sup_{c_1,\ldots,c_T} \left[\sup_{x^*} \Big(\sum_{t=1}^{T} \mathcal{U}(c_t, x^*(c_t)) \Big) - \sum_{t=1}^{T} \mathcal{U}(c_t, x_t(c_t)) \right].$$

Simply put, the regret of an algorithm is the worst-case difference between its achieved utility and the optimal utility achieved by a fixed decision rule. Many variants of the above notion of regret exist, including those that replace the left-most supremum with an expectation and those which allow the utility function to change over time. In Example 5, if a suboptimal decision is made on the first context c_1 (i.e., $\mathcal{U}(c_1, x^*(c_1)) - \mathcal{U}(c_1, x_1(c_1)) = \delta > 0$), then by choosing $c_1 = \cdots = c_T$, we see that the regret up to time T is at least $T\delta = \Omega(T)$, which indicates poor performance. This asymptotic lower bound holds when contexts are generated randomly as well, as long as the distribution over C is reasonable.

There are two ways around this predicament in Example 5: first, one can *reduce the memory* of the fairness constraint, thus allowing for more flexibility in adjusting decisions over time. Second, one can *relax the fairness constraint*. Weakening the fairness constraint may allow for learning, even in the full memory framework.

Example 7

Recall from Example 5 that enforcing individual fairness with full memory can lead to poor performance. In this example, we show how such notions of comparative fairness can be *temporally relaxed*, potentially avoiding this issue. In particular, suppose we have a finite context space $C = \{c_1, \ldots, c_N\}$ and are tasked with assigning probabilities of selection in $\mathcal{X} = [0, 1]$ for applicant screening.

If we were to impose individual fairness with full memory, as in Example 5 , then the decision $x_t(c_i)$ for context c_i at time t is constrained from *above and below* by decisions made at times $1, \ldots, t-1$. However, a job applicant will only feel mistreated if their decision is worse than expected with respect to decisions made in the past (i.e., if they are rejected and a similar applicant was selected in the past). Motivated by this, a one-sided relaxation of comparative fairness called *fairness at the time of decision* (FTD) has emerged in the literature [20]. In particular, given slacks $s(i, j) \in \mathbb{R}$ for $i, j \in [N]$, the constraints imposed at time t are

$$x_t(c_i) \geq x_{t'}(c_j) - s(i, j) \quad \text{for all } t' \leq t \text{ and } i, j \in [N].$$

Importantly, these constraints allow decisions to become more conducive to applicants (i.e., to increase) over time, while ensuring that no context is treated unfairly *relative to decisions made in the past*. On the other hand, since decisions are allowed to increase arbitrarily under this constraint, a decision made on context c_i at time t may be viewed as unfair compared to a decision made at time $t + 10$. The benefit of this weaker form of comparative fairness is that this ability to increase decisions over time allows for learning in some scenarios where the unrelaxed constraint would not, as we illustrate in the next section.

As discussed above, there are important normative questions about the meaning of fairness in dynamic decision-making and technical questions about the trade-offs between temporal notions of fairness and performance. In a given decision-making setting, how can one balance memory and stringency of a fairness constraint with the flexibility required to learn a good decision? These questions pose technically new challenges in online optimization and require new design tools and techniques. In the next two sections, we illustrate techniques and challenges in the design of temporally fair algorithms.

4 Example I: Partial Memory Comparative Fairness

In this section, we discuss a partial memory comparative fairness constraint proposed for the bandit online learning framework by Heidari and Krause in 2018 [11]. In their setting, a sequence of contexts $c_1, c_2, \ldots \in C$ is observed one by one over time. Upon observing a context c_t, a decision $x_t(c_t) \in \mathcal{X} := [0, 1]$ is made, and after the decision is made, a true label y_t is observed. The authors chose comparative fairness as a goal; in particular, for a given metric d on C, their goal is to ensure that

$$|x_t(c_t) - x_{t'}(c_{t'})| \leq d(c_t, c_{t'}) + s \tag{4}$$

for all time periods t, t' and slack $s \geq 0$.

However, the authors note that enforcing the above full memory fairness constraint prevents learning. A bad decision in the first context c_1 can prevent good decisions for the remaining time periods, thus leading to large errors. To get around this issue, the authors instead imposed a partial memory version of the above constraint, where the constraint is only enforced in a *sliding window*. Specifically, they refer to a sequence $\{(c_t, x_t(c_t))\}_{t=1}^{T}$ of contexts and decisions as (s, m)-*consistent* if (4) holds whenever $|t - t'| \leq m$.

The authors opt for a post-processing approach to making (s, m)-consistent decisions, called Consistently Follow the Leader (CFTL). In particular, given context c_t at time t, a black box algorithm chooses a decision x, and this decision is transformed into an (s, m)-consistent one. Letting I denote the set of consistent decisions for c_t at time t, the authors show that I is a compact, nonempty interval, assuming that the prior decisions were (s, m)-consistent. To obtain a consistent decision at time t, they simply project the black box decision onto I; i.e., $x_t(c_t) = \text{Proj}_I(x)$.

The performance of CFTL, of course, depends on the black box algorithm. The authors assess the quality of CFTL using a tailored notion of learnability: a hypothesis class is called *(s, m)-CS learnable*[2] if there is an algorithm which (1) produces (s, m)-consistent decisions, and (2) after N rounds, chooses an almost-optimal hypothesis with probability at least $1 - \delta$. Using CFTL, the authors show that if the hypothesis class has PAC sample complexity N, then it has (s, m)-CS sample complexity of at most $N + \frac{m}{s}$. More precisely, assuming that the black box algorithm is PAC-learning with sample size N, CFTL is (s, m)-CS-learning with sample size $N + \frac{m}{s}$. Moreover, the authors show that this dependence on m and s is tight.

These results are interesting as they illuminate some trade-offs between memory of fairness constraints and learnability. The work of Heidari and Krause [11], which shows a linear cost in the memory parameter to the sample complexity, adds to the literature identifying and quantifying costs of temporal constraints [5, 13]. In contrast, in the next section, we discuss a setting in which imposing a full memory constraint does not increase achievable regret.

5 Example II: Full Memory Relaxed Comparative Fairness

Recall that in Sect. 4, we discussed a partial memory enforcement of approximate comparative fairness. In this section, we explore a variant of this constraint which (1) is full memory, (2) allows for different and asymmetric slacks for different pairs of contexts, and (3) only constrains decisions *from below* based on decisions made in the past. Specifically, the constraint we consider is *fairness at the time of decision* (FTD), as introduced in Example 7, with the modification that $s(i, i) = 0$ for every context c_i. This essentially imposes the following constraints:

1. **Coordinate-wise monotonicity**. Decisions must increase for each group over time; i.e., for every context c_i, $x_1(c_i) \leq x_2(c_i) \leq \cdots \leq x_T(c_i)$, where $x_t(c_i)$ is the decision given to context c_i at time t; and
2. **Cross-coordinate constraints**. For any pair of context (c_i, c_j) and time periods $t' \leq t$, $x_t(c_i) \geq x_{t'}(c_j) - s(i, j)$.

Essentially, this constraint requires that decisions increase over time for every context and the the decision $x_t(c_i)$ given to context c_i at time t is large enough with respect to decisions made in the past on other contexts. Note that this constraint and the one discussed in Sect. 4 are not directly comparable: neither is a generalization of the other.

We explore this new constraint in the setting of stochastic convex optimization. Consider a simple setting where a principal chooses decisions in $\mathcal{X} = [x_{\min}, x_{\max}] \subseteq \mathbb{R}^{\geq 0}$ for each of N groups (i.e., N contexts) repeatedly over T periods. A decision

[2] CS is short for "consistent sequential."

x for group i leads to a cost $f_i(x)$, where f_i is assumed to be a β-smooth and α-strongly convex function for each i. f_i is not known a priori to the principal. Upon choosing a decision $x_t(c_i)$ at time t for group c_i, the principal observes the noisy feedback $f_i(x_t(c_i)) + \varepsilon_{i,t}$, where $(\varepsilon_{i,t})_t$ is a sequence of independent zero-mean sub-Gaussian random variables. The goal of the principal is to choose decisions to minimize $\sum_{i=1}^{N} f_i$ over time while ensuring that the decisions satisfy FTD.

Without any constraints, it is well known that $\Omega(\sqrt{T})$ regret is inevitable: this lower bound for the case of smooth and strongly convex functions is due to [21]. A near-optimal algorithm for this case has been designed by [1]. In the one-dimensional case, their approach is the most related to the golden-section search procedure of [14]: it iteratively uses three-point function evaluations to "zoom in" to the optimum, by eliminating a point and sampling a new point in each round. Its mechanics render it infeasible to implement it in a fashion that respects the monotonicity of decisions. Another algorithm achieving this guarantee is due to [10] and is based on gradient descent using a one-point gradient estimate constructed by sampling uniformly in a ball around the current point. This key idea recurringly appears in several works on convex optimization with bandit feedback (e.g., [7, 10, 22]). However, due to the randomness in the direction chosen to estimate the gradient, such an approach does not satisfy the FTD constraints. It is thus unclear if the optimal regret rate of $\widetilde{O}(\sqrt{T})$ can be obtained while satisfying the FTD constraint. This question was recently answered in the affirmative for $N = 1, 2$ [20]. However, for ease of exposition, we introduce a weaker statement for general N.

Theorem 1 (Lagged Gradient Descent regret bound [20]) *Suppose that the minimizer* $\mathbf{x}^* = \arg\min_{\mathbf{x}\in\mathbb{R}^2} \sum_{i=1}^{N} f_i(x_i)$ *is such that* $x_i^* > x_{\min}$ *for each* i. *Then there exists an algorithm for choosing decisions for* N *groups over time that satisfies FTD and that attains a regret of* $\widetilde{O}(N^3 T^{2/3})$ *under noisy bandit feedback assuming that* $f(x_1, \ldots, x_N) = \sum_{i=1}^{N} f_i(x_i)$ *is* β*-smooth and* α*-strongly convex.*

There are several algorithmic ideas in achieving the result of Theorem 1. Discussing all of them is beyond the scope of this article and so we focus on the key ones in Sects. 5.1–5.3. The reader can refer to [20] for details.

5.1 Algorithm Design for a Single Context

Let's first focus on the problem of optimizing a single-dimensional function (i.e., $N = 1$) using bandit feedback *while ensuring that the decisions are monotonically increasing*. Starting from $x_{\min}$, the decisions need to increase to the unknown optimum at a sufficient pace to ensure low regret. However, hastiness is associated with an increased risk of overshooting the optimum, which would lead to high regret since backtracking is not allowed.

The main idea for addressing this trade-off is to tailor the degree of caution (i.e., the speed of approach) to the local gradient. Indeed, if we had access to gradient feedback, then it would be easy to show that the standard gradient descent dynamics

$$x_{t+1} = x_t - \eta \nabla f(x_t),$$

beginning with $x_0 = x_{\min}$ *monotonically* converge to the optimum at an exponential rate *while never overshooting the optimum*, assuming that the step-size η is chosen appropriately as a function of β (the smoothness parameter). However, to execute such a procedure with noisy bandit feedback, an estimate of the gradient at x_t must be constructed. We can do so by sampling the function repeatedly at two points x_t and $x_t - \delta$ separated by some lag $\delta > 0$ (we first sample at $x_t - \delta$ and then at x_t to ensure monotonicity). Overshooting can then be avoided by moving from the lagged point, i.e.,

$$x_{t+1} = x_t - \delta - \eta g_t$$

assuming that g_t is a gradient estimate that satisfies $\nabla f(x_t - \delta) \leq g_t \leq \nabla f(x_t)$. If x_{t+1} is smaller than x_t, then we stop the procedure to ensure monotonicity. The following lemma characterizes the sample complexity of calculating g_t.

Lemma 1 (Sandwich Lemma [20]) *Let* $f : \mathbf{R} \to \mathbf{R}$ *be an* α*-strongly convex function. Let* $x < y$ *and let* $\delta = y - x$*. Fix* $p \in (0, 1)$*, and define* $\overline{f}(x)$*,* $\overline{f}(y)$ *to be the averages of* $\Theta(\frac{\log \frac{1}{p}}{\alpha^2 \delta^4})$ *samples at* x *and* y*, respectively. Then the estimated secant* $g = \frac{\overline{f}(y) - \overline{f}(x) + \frac{\alpha\delta^2}{4}}{\delta}$ *satisfies* $\nabla f(x) \leq g \leq \nabla f(y)$*, with probability at least* $(1 - p)^2$.

In other words, to estimate the gradient of a strongly convex function between two points separated by δ, it is sufficient to sample these points $O(1/\delta^4)$ times (we can also argue that this is necessary). This means that δ cannot be too small, otherwise excessive regret would be incurred in the gradient estimation process. At the same time, if δ is too large in relation to g_t then the procedure may stop prematurely and far from the optimum, resulting in high regret. We can show that one can choose an appropriate value of δ that balances the estimation regret and the stopping regret to yield a $\widetilde{O}(T^{2/3})$ regret monotone algorithm.

5.2 *Algorithm Design for N Contexts*

Now to address the multi-group case, we need to additionally address the cross-coordinate constraints, i.e., $x_t(c_i) \geq x_{t'}(c_j) - s(i, j)$ for all $1 \leq t' \leq t \leq T$ and $i \neq j \in [N]$. Our overall approach is more simply described as a continuous-time procedure in the case where we have access to perfect gradient feedback, i.e., $\nabla f_i(x_t(c_i))$:

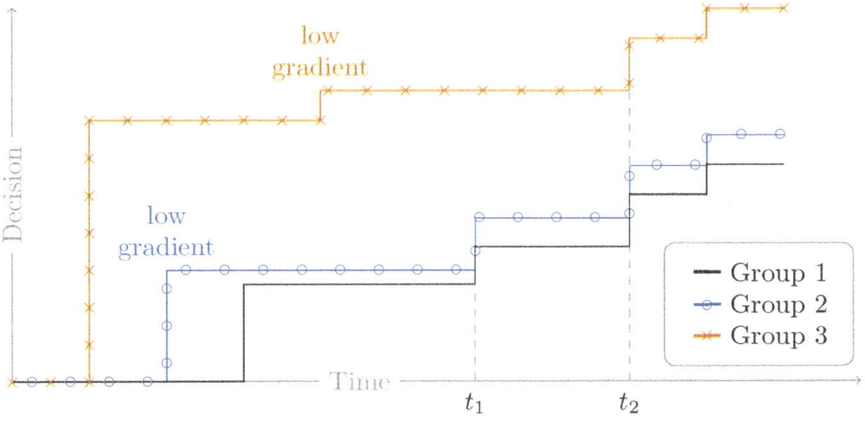

Fig. 1 An illustration of the multi-context approach described above with three contexts (i.e., three groups). Groups 1 and 2 are combined into a cluster at time t_1, and that cluster is combined with Group 3 at time t_2. Note that only one cluster can move at a time

Continuous-Time Approach
Assign the ith context to a cluster $\{i\}$, for $i \in [N]$, and repeat the following until all clusters are at their optima:

1. While there exists a cluster j which is not at its optimum and not at the boundary of the feasible region, increase its decision until it either hits its optimum or hits a constraint.
2. If there is a cluster C which is not at its optimum, find a cluster D which is constraining the movement of C. Replace both clusters with the new cluster $C \cup D$. The function corresponding to $C \cup D$ that will be optimized is the sum of the functions corresponding to C and D.

See Fig. 1 for an illustration of this procedure.

The challenge then is to convert this process into a practical discrete-time procedure with only noisy bandit feedback on each dimension, *while ensuring optimal overall regret*. As discussed in Sect. 5.1, for the single-context setting, gradient descent movements can be made in discrete time using samples made at the current point $x_t(c_i)$ and the lagged point $x_t(c_i) - \delta$. The additional challenge in this multi-context setting is deciding when to combine clusters. Using a lag size of δ, we can only detect optimality up to an error of δ^2, and these errors can accumulate as clusters are combined. This accumulation of errors ultimately produces a cubic dependence on the number of contexts, as shown in Theorem 1. We refer the reader to [20] for a more detailed description of the algorithm and analysis.

5.3 Improvements and Open Directions

Note that the $\widetilde{O}(N^3 T^{2/3})$ regret bound presented in Theorem 1 leaves a gap with the best-known lower bound of $O(\sqrt{T})$. In [20], the authors show that this gap can be closed for $N = 1, 2$. The bottleneck with the single-context fixed-lag approach presented above is that progress can only be made if the jump size $-\eta g_t - \delta$ is non-negative; so, once the gradient is of order δ (and thus instantaneous error is of order δ^2), no more changes can be made.

To get around this issue, we adaptively tailor the lag size to the local gradient estimate. In particular, if we find that the algorithm has stopped moving for a particular lag size δ, then we halve the value to $\delta/2$ and attempt to keep moving, halving the value further as necessary. The benefit of this approach is that smaller values of the lag size δ, which result in a high sampling rate for gradient estimation, are utilized only when the decisions are close to the optimum where they result in low regret. One complication of this approach is that the lag size δ for an iterate x_t must be known before sampling at x_t, since sampling at the lagged iterate $x_t - \delta$ after sampling at x_t would violate monotonicity. We tackle this issue by constructing interim gradient estimates to search for the right lag size before sampling x_t. The details of this design are beyond the scope of this article. Overall, we can show that incorporating adaptive lags to our single-context and two-context methods attains the optimal $\widetilde{O}(\sqrt{T})$ regret guarantee [20].

There are a number of open questions stemming from this work, including:

1. **Relaxing regularity assumptions**. The assumption of smoothness (which allowed the elimination of overshooting) and strong convexity (which is crucial for the sandwich lemma) appears to be crucial for the result of Theorem 1. Jia et al. [13] and Chen [5] have recently considered the problem of stochastic optimization of an unknown single-dimensional Lipschitz unimodal function under bandit feedback, and they have established that enforcing monotonicity of decisions results in the optimal achievable regret increasing from $\widetilde{\Theta}(T^{2/3})$ to $\widetilde{\Theta}(T^{3/4})$. So, it appears that some form of regularity of the functions is necessary to ensure that there is no impact of the FTD constraint on the optimal regret (at least up to logarithmic terms). In fact, it appears that smoothness is necessary upon considering the worst-case examples of [13] and [5]. Intuitively, without smoothness (or something similar in its place), it is not possible to anticipate the optimum as one monotonically changes the decisions, and thus excessive overshooting of the optimum is inevitable in the worst-case. However, it is an open question whether strong convexity can be relaxed.
2. **FTD with multiplicative slacks**. In many scenarios, multiplicative allowable disparities in treatment could be more appropriate than additive disparities. For example, a firm may want to price an item while ensuring that the price disparity between the youth and the general population is not more than 25%. In particular, consider slack functions $\bar{s}(\cdot,\cdot) : \mathcal{I} \times \mathcal{I} \to \mathbb{R}^{\geq 0}$ defined on ordered pairs of groups, and consider the following definition of FTD with multiplicative slacks.

Definition 1 (*FTD with Multiplicative Slacks*) We say that a decision vector $x \in \mathcal{X}^N$ satisfies *fairness at the time of decision (FTD) with multiplicative slacks* if the following inequalities hold:

$$x_i \geq \bar{s}(i, j)x_j \text{ for all } i, j \in \mathcal{I}. \tag{5}$$

It would be interesting to extend our algorithm design to satisfy an FTD extension of this kind. One setting where our current results extend is the case where there is a function $m : \mathcal{I} \to \mathbb{R}^{>0}$ such that $\bar{s}(i, j) = m(j)/m(i)$. In this case, (5) effectively requires that $x_i m_i \geq x_j m_j$. In this case, one can redefine the decision space for group i as $y_i = x_i m_i$. Our algorithms then readily apply to optimizing y in this setting.

3. **Slack-monotonicity**. In our analysis of fairness at the time of decision with additive slacks, we assumed that $s(i, i) = 0$. This resulted in the temporal monotonicity constraint $x_t(c_i) \geq x_{t'}(c_i) - s(i, i) = x_{t'}(c_i)$ for all i and $t \geq t'$. However, in many scenarios, some changes to within-group decisions across time may be acceptable. In other words, we may have $s(i, i) > 0$ for all i, so that we effectively require that $x_t(c_i)$ is at most $s(i, i)$ lower than the highest decision seen by group i until time t, thus allowing for a limited amount of backtracking in decisions (similarly, with multiplicative slacks, we may have that $s(i, i) < 1$). While it is interesting that this type of slack in the monotonicity constraint isn't necessary for our setting to attain the unconstrained optimal near-regret rate of $\widetilde{O}(\sqrt{T})$, it may nevertheless simplify the algorithm design. Moreover, it may provide crucial flexibility in attaining near-optimal regret rates in harder settings, such as when the functions are only known to be unimodal.
4. **Extensions to online convex optimization**. It would be interesting to consider algorithm design for satisfying FTD in the general context of online convex optimization, which models scenarios where the utility function $\mathcal{U}$, though convex, changes over time [9]. In such scenarios, even with perfect gradient feedback (since the utility functions $\mathcal{U}_t$ are assumed to be known in each period), it would be interesting to characterize conditions on the arriving functions under which sublinear regret rates can be attained while satisfying FTD.
5. **Time decay**. There is room for generalizing or adapting FTD to better fit specific objectives. For example, one may wish to incorporate only partial memory (e.g., a time-decay element), which can be achieved by having the slack functions s be dependent on the difference in the times at which the decisions were received by the two groups. In particular, we may have that $x_t(c_i) \geq x_{t'}(c_i) - s(i, i, t - t')$ for all $t \geq t'$, where the slack function now also takes the time difference $t - t'$ as input. If s is assumed to be increasing in the time difference, this would allow for greater changes in decisions over time.

6 Connections with Law and Policy

Finally, we note that temporal considerations can be required or suggested for compliance with laws or policies. For example, applicant screeners often ensure approximate demographic parity for the purpose of avoiding disparate impact litigation [17], and this constraint constrains current decisions by prior decisions. So-called "mandated progress" legislation, such as affirmative action—which has been adopted and, in some cases, mandated in the U.S., Canada, and France—can similarly be thought of as a temporal constraint in which goals must be set and a current decision is thus constrained by past decisions [15]. In terms of loan granting, Wells Fargo was recently sued for racial discrimination in mortgage lending, including offering different average interest rates to Black applicants than white applicants over a period of time [23], which again points to the potential for the use of temporal fairness constraints as a preventative measure for avoiding litigation. Sometimes these constraints can be enforced due to the necessity of protecting consumer rights in the prevalent societal context; e.g., price gouging laws may dictate bounded increases in decisions during the duration of the pandemic (*McQueen and Ballinger v. Amazon.com*). In general, the case law surrounding algorithmic approaches to ensuring fairness is not yet well-developed, thus leaving many questions open. However, we believe that it is important to get ahead of such restrictions and to understand the limitations of algorithms in order to provide guidance to legal scholars on the possibilities.

References

1. Agarwal A, Foster DP, Hsu D, Kakade SM, Rakhlin A (2013) Stochastic convex optimization with bandit feedback. SIAM J Optim 23(1):213–240
2. Baker W, Kiewell D, Winkler G (2014) Using big data to make better pricing decisions. McKinsey Analysis
3. Bansal S, Srivastava A, Arora A (2017) Topic modeling driven content based jobs recommendation engine for recruitment industry. Procedia Comput Sci 122:865–872
4. Barocas S, Hardt M, Narayanan A (2019) Fairness and machine learning: limitations and opportunities. http://www.fairmlbook.org
5. Chen N (2021) Multi-armed bandit requiring monotone arm sequences. arXiv:2106.03790
6. Chouldechova A, Roth A (2018) The frontiers of fairness in machine learning. arXiv:1810.08810
7. Flaxman AD, Kalai AT, McMahan HB (2005) Online convex optimization in the bandit setting: gradient descent without a gradient. In: Proceedings of the sixteenth annual ACM-SIAM symposium on Discrete algorithms, pp 385–394
8. Gupta S, Kamble V (2021) Individual fairness in hindsight. J Mach Learn Res 22(144):1–35
9. Hazan E (2016) Introduction to online convex optimization. Found Trends Optim 2(3–4):157–325
10. Hazan E, Levy KY (2014) Bandit convex optimization: towards tight bounds. In: NIPS, pp 784–792
11. Heidari H, Krause A (2018) Preventing disparate treatment in sequential decision making. In: Proceedings of the 27th international joint conference on artificial intelligence, pp 2248–2254
12. Jagtiani J, Lemieux C (2019) The roles of alternative data and machine learning in fintech lending: evidence from the LendingClub consumer platform. Financ Manag 48(4):1009–1029

13. Jia S, Li A, Ravi R (2021) Markdown pricing under unknown demand. Available at SSRN 3861379
14. Kiefer J (1953) Sequential minimax search for a maximum. Proc Am Math Soc 4(3):502–506
15. Klarsfeld A, Cachat-Rosset G (2021) Equality of treatment, opportunity, and outcomes: mapping the law. In: Oxford research encyclopedia of business and management
16. Panesar A (2019) Machine learning and AI for healthcare. Springer
17. Raghavan M, Barocas S, Kleinberg J, Levy K (2020) Mitigating bias in algorithmic hiring: Evaluating claims and practices. In: Proceedings of the 2020 conference on fairness, accountability, and transparency, pp 469–481
18. Rigano C (2019) Using artificial intelligence to address criminal justice needs. Natl Inst Justice J 280:1–10
19. Salem J, Gupta S (2023) Secretary problems with biased evaluations using partial ordinal information. Manage Sci
20. Salem J, Gupta S, Kamble V (2022) Algorithmic challenges in ensuring fairness at the time of decision. arXiv:2103.09287
21. Shamir O (2013) On the complexity of bandit and derivative-free stochastic convex optimization. In: Conference on learning theory, PMLR, pp 3–24
22. Spall JC et al (1992) Multivariate stochastic approximation using a simultaneous perturbation gradient approximation. IEEE Trans Autom Control 37(3):332–341
23. Waters T (2022) Wells fargo bank sued for race discrimination in mortgage lending practices. https://www.usatoday.com/story/money/2022/04/26/wells-fargo-being-sued-discriminating-against-black-borrowers/7451521001/. Published 26 Apr 2022. Accessed 7 Jun 2022

No AI After Auschwitz? Bridging AI and Memory Ethics in the Context of Information Retrieval of Genocide-Related Information

Mykola Makhortykh

Abstract The growing application of artificial intelligence (AI) in the field of information retrieval (IR) affects different domains, including cultural heritage. By facilitating organisation and retrieval of large volumes of heritage-related content, AI-driven IR systems inform users about a broad range of historical phenomena, including genocides (e.g. the Holocaust). However, it is currently unclear to what degree IR systems are capable of dealing with multiple ethical challenges associated with the curation of genocide-related information. To address this question, this chapter provides an overview of ethical challenges associated with the human curation of genocide-related information using a three-part framework inspired by Belmont criteria (i.e. curation challenges associated with respect for individuals, beneficence and justice/fairness). Then, the chapter discusses to what degree the above-mentioned challenges are applicable to the ways in which AI-driven IR systems deal with genocide-related information and what can be the potential ways of bridging AI and memory ethics in this context.

1 Introduction

Information retrieval (IR) is one of the computer science fields that is closely connected to the developments in the domain of artificial intelligence (AI). Defined as the process of selecting items that are deemed relevant for the user information needs based on the user input [1], IR has been argued to be a particularly promising area of applying AI [2, 3]. The integration of AI can benefit different aspects of IR, ranging from knowledge representation to content indexing and matching [6] to relevance modelling [3, 4]. Consequently, there is a long history of research on AI-driven IR applications, starting with rule-based approaches in the 1980s [2] and ending with the neutral network-based approaches discussed in the 2020s [4].

M. Makhortykh (✉)
University of Bern, Fabrikstrasse 8, Bern, Switzerland
e-mail: mykola.makhortykh@unibe.ch

A. Mukherjee et al. (eds.), *Ethics in Artificial Intelligence: Bias, Fairness and Beyond*, Studies in Computational Intelligence 1123,
https://doi.org/10.1007/978-981-99-7184-8_4

The importance of AI-driven IR systems has been increasing due to the growth in the amount of information available online. Often referred to as an information overload [5], this phenomenon prompted the need for advanced IR mechanisms for satisfying individual information needs which are capable of not only processing the large volumes of available information, but also recognising the diverse spectrum of user needs and in some cases predicting these needs. Such mechanisms demonstrated their usefulness in multiple domains ranging from healthcare [7, 8] to journalism [9, 10] to e-commerce [11, 12].

This chapter focuses on one particular domain in which AI-driven IR systems are increasingly employed which is cultural heritage. By facilitating the organisation of heritage-related content both within heritage institutions (e.g. archives [13] or museums [14]), and commercial platforms (e.g. web search engines [15] or social media news feeds [16]), AI-driven IR systems help their users become informed about a broad range of historical phenomena, including genocides such as the Holocaust or Rwanda genocide. Under the condition of a high degree of autonomy, these systems become non-human curators of genocide-related information which shape how individuals and societies are informed about the past and present atrocities.

Despite the importance of IR systems for curating information about historical and recent genocides, there are multiple concerns about their potential impact on genocide remembrance. The usual concerns about the lack of transparency of AI-driven IR systems are further amplified by the possibility of such non-transparency facilitating manipulations of IR systems which can potentially interfere with the moral obligations of safeguarding the dignity of genocide victims [17]. Furthermore, such manipulations can facilitate the instrumentalisation of memories about past violence which can be used for justifying the present stigmatisation as in the case of the Rohingya persecutions in Myanmar [18] or the Russian-Ukrainian war [19].

Besides the above-mentioned concerns about the use of IR for curating genocide-related information, there are also other ethical challenges which have for long been discussed in the context of human curation of historical information such as the importance of protecting the privacy of individuals [20] or preventing unfair practices of information curation [21]. However, despite the growing body of work concerning the ethics of human curation of genocide-related information [22–24], the capabilities of IR systems to deal with complex ethical issues arising in the context of genocide-related information as well as the perspectives of bridging memory ethics and IR design currently remain under-investigated.

To address this gap, the chapter aims to examine whether the concerns about the human curation of genocide-related information are applicable to AI-driven IR systems and how these concerns can potentially be addressed. For this aim, it provides a short overview of the current applications of IR systems in the context of genocide-related information, followed by a discussion of the ethical challenges of its human curation using a three-part framework inspired by Belmont criteria (i.e. respect for individuals, beneficence and justice/fairness). Finally, the chapter discusses to what degree these challenges are applicable to AI-driven IR systems and what can be the potential ways of bridging AI and memory ethics in this context.

2 AI-driven IR Systems and Genocide-Related Information

The digitisation of historical collections together with the production of new digital-born materials dealing with information about genocides (e.g. audiovisual tributes to the Holocaust [25, 26]) prompted the growing use of IR systems within heritage institutions. Many of these systems partially reproduce or enhance the traditional curation practices used by archives or museums, but in some cases, IR systems substantially transform the scale and functionality of these practices.

For instance, Liew [14] examined how IR systems facilitate the exploration of collections, including the ones dealing with the Holocaust (e.g. by enabling keyword/phrase search and the use of wildcard operators). Schenkolewski-Kroll and Tractinsky [13] discussed the relationship between IR systems and authority lists in the context of Holocaust materials in the Israeli archives. Daelen [27] looked at the possibilities of using IR to enrich the inventory of collections related to the Holocaust and connect archives and users in the context of European Holocaust Research Infrastructure; similarly, Carter et al. [28] discussed the potential of AI-driven IR solutions for facilitating exploration of primary sources in the context of the Holocaust by proving new possibilities for user interaction with the Morgenthau Diaries.

In addition to IR systems enhancing traditional curation practices, there are also examples of more innovative applications of these systems within heritage institutions. One example is the use of three-dimensional visualisation of Holocaust survivors retrieving audio recordings of the survivors' earlier comments in response to the user input. Sometimes referred to as holograms [29] or social robots [30], these systems are used by several Holocaust memory initiatives (e.g. New Dimensions of Testimony or Forever Project) and enable new possibilities to retrieve testimonies through the simulation of human-to-human conversation.

Similar to other experimental proposals concerning the use of AI-driven IR systems in the context of genocide remembrance (e.g. the concept of personalised virtual reality-enhanced interaction with information about the Holocaust for Babyn Yar memorial [31]), the use of social robots as a form of curation of genocide-related information has attracted not only praise, but also criticism. For instance, Walden [29] noted that novel approaches for the use of IR (e.g. in the form of Holocaust survivors' holograms) do not necessarily meet the expectations about reactualisation of the past for the audience, whereas Alexander [32] noted that some of these novel approaches require not only historical knowledge, but also media literacy, thus risking to make information less accessible for certain groups.

Not only heritage institutions, but also commercial platforms are increasingly relying on IR systems for curating genocide-related information. The availability of digital content related to genocides coming both from the institutional (e.g. Holocaust museums [33]) as well as non-institutional entities such as online influencers [34] or artists [25], resulted in the growing presence of genocide-related information on the online platforms. These platforms range from social media sites, such as Instagram [34] or TikTok [35] to web search engines such as Google [36] to commerce-related platforms (e.g. TripAdvisor [37]).

Under these circumstances, IR systems become a crucial element of genocide-related information curation, in particular, considering that commercial platforms often lack human curation expertise in this specific domain that differentiates them from heritage institutions. However, the implications of IR-driven curation currently remain unclear. Makhortykh et al. [36] examined how six web search engines curate visual information about the Holocaust and observed substantial differences in what aspects of the Holocaust are prioritised by the individual engines. Devon and Tobias-Hartmann [35] discussed the impact of the TikTok algorithm on the treatment of user-generated content, including content dealign with Holocaust denial, and found the tendency of the algorithm to suppress certain forms of resistance to antisemitic ideologies. Finally, Kansteiner [38] looked at how Holocaust institutions use IR systems associated with commercial social media sites (e.g. Facebook) and found the varying degrees of visibility of specific types of genocide-related content received.

Unsurprisingly, the use of AI-driven IR systems by commercial platforms for curating information about genocides also raised a number of concerns. In addition to the general critique of it undermining the gatekeeping functions of heritage institutions [39], studies suggest that IR systems used by platforms can promote factually incorrect or denialist content [36, 40]. Another concern relates to the possibility of commercial platforms' IR systems resulting in unequal treatment of information about different aspects of specific genocides (e.g. in terms of prioritising content coming from a few Holocaust sites while omitting the other ones [36]).

3 Memory Ethics and Human Curation of Genocide-Related Information

The major challenge of bridging AI and memory ethics in the context of genocide-related information curation deals with the multiple forms the curation might take. There are many approaches to human curation of such information ranging from the one happening in heritage-focused environments, for instance, museums or archives, to wider public-focused environments, such as mass media.

The multiplicity of forms of curation prompts the importance of identifying which of them are particularly applicable to the discussion of AI-driven IR systems. While it can be debated, the argument can be made that human curation in archives is the closest in its nature. Both archives and IR systems determine what content is made visible to the public and what content remains hidden [23]: in the case of archives, these decisions are implemented by providing or not providing physical access to the collections, whereas in the case of IR systems, some outputs in response to user queries can be filtered out or down ranked.

Human curation of genocide-related information in the archival context has to deal with multiple ethical challenges [21, 24]. Some of these challenges are applicable to archival research in general, for instance, the potential damages to individual privacy [23]. However, other challenges are more specific to the case of genocide and include, for instance, the possibility of using archives to subjugate knowledge about the past atrocities [41] or impeding the processing of genocide-related trauma by encouraging specific types of testimonies and silencing the others [42]. The need to address these challenges stimulates the discussion of how these ethical challenges can be addressed.

One of the common reference points in the discussion of ethics regarding human curation of archival information is Belmont criteria. Introduced at the end of the 1970s to provide guidelines for research involving human subjects, Belmont's criteria focus on three ethical principles: respect for individuals, beneficence, and justice (sometimes also referred to as fairness [24]). The recommendations of Belmont criteria generally suggest that "informed consent be sought, that benefits and risks be evaluated, and the selection, representation, and the burden of participation be fair and equitable" [43, p. 139].

A number of studies have critically interrogated to what degree Belmont criteria are applicable for the archival research [22, 24]. Some studies argued that Belmont criteria are not applicable to archival research, because it is fundamentally different from the other disciplines working with human subjects [23], whereas others (e.g. [24]) suggested that the criteria are focused primarily on preventing potential damage for the living subjects. However, in the case of genocide-centred research, many subjects are already dead that makes it hardly possible to obtain their consent for being involved in the research and the different set of risks/threats (e.g. potential damage to posthumous dignity of victims [17]) which have implications for the beneficence and justice criteria. Under these conditions, direct application of Belmont criteria to archival research dealing with genocides may undermine the ethical mandate of the genocide-focused scholarship [22].

Despite the above-mentioned drawbacks, it can be argued that Belmont criteria are still applicable for identifying the ethical challenges involved in human curation of archival information about genocides. Specifically, this chapter proposes to apply a three-part framework using Belmont criteria—i.e. respect for individuals, beneficence and justice/fairness—to group together potential ethical challenges associated with genocide-related information curation. The rest of the section is devoted to the discussion of the individual challenges associated with each of the three criteria.

In the case of respect for individuals, it is possible to identify three major ethical challenges related to the human curation of genocide-related information: consent, double vision, and privacy. The first of these challenges—i.e. the need to acquire consent—is common for curation of information coming from the human subjects in other contexts. However, in the case of genocide or other forms of mass violence, making sure that the consent is acquired becomes a much harder task. In some cases, the difficulties can be due to substantial risks for witnesses or victims preventing them from voluntarily sharing information [44] or evidence being produced against the will of the victims [45].

Furthermore, the digitisation of genocide-related information raises additional questions such as, for instance, whether the consent given for the generation of analogue materials also automatically applies to their digitalisation and whether the difference between analogue and digital public access has implications for the consent to make information about the genocide publicly available [20]. These questions are particularly applicable for the historical instances of genocide (e.g. the Holocaust), where materials (e.g. testimonies) were produced in certain formats which has since then become outdated, so preserving them in the original format is both non-sustainable and ineffective from the point of view of communicating information about the genocide.

Another challenge of human curation relates to the problem of the double vision, which relates to the transformation of genocide victims into objects (and not subjects) of research due to the distancing involved in the process of data collection and analysis [46]. Originally discussed in the context of processing analogue materials [22, 46], the problem of potential dehumanisation and depersonalisation of victims is amplified by the shift towards digital collections enabling new possibilities for "anonymizing, numbering, and classifying" [24, p. 531] experiences of genocide victims as well as "converting humans into numbers" [47, p. 322].

One more aspect of respect for individuals concerns the matters of privacy. Archives in general can be damaging to the reputation of individuals whose information is disclosed without their consent [20, 23]. However, in the case of genocide, in particular its recent instances, privacy can be a matter of life and death, for instance, when either perpetrators or victims want to take revenge in their hands. At the same time, the profound anonymisation of genocide-related records has been criticised for its potential for erasing the voice of victims [22] which in some way is similar to the purpose of the genocidal actions aiming to erase any traces of victims.

For the beneficence of the human curation of genocide-related information, it is possible to identify two major challenges: the problem of representation and the possibility of distortion/manipulation. The former challenge relates to the argument that because genocides are instances of unprecedented violence, any attempt of their representation (e.g. via certain modes of storytelling or documentation [30, 48]) is inadequate. Hence, the absence of representation might be a "more accurate or truthful or morally responsive" [49, p. 71] way of dealing with genocide-related information.

The second challenge concerns the possibility of information about genocide being distorted or manipulated. The forms of distortion of historical information can vary broadly; some examples include de-contextualisation of historical phenomena [20], denial or justification of the past crimes [50] or the use of references to past suffering or injustice for stigmatising specific social groups in the present [19]. In the case of genocide-related information, such forms of distortion are particularly concerning both due to the ethical obligations of protecting the memory of victims and the strong affective potential of information about past injustices which can be used to incite violence in the present [26].

Finally, the justice/fairness of human curation concerns two interrelated aspects: the politicisation of curation and the unequal treatment of specific types of genocide-related information. One of them is the politicisation of archives that has implications for what information about the past atrocities is available and how it is communicated to the public [24, 47]. The transmission of the matters of curation of genocide-related information to the realm of politics, might not only downplay the importance of ethical obligations associated with it, but also facilitate instrumentalisation of genocide memory for immediate political gains. Such instrumentalisation can lead to genocide-related information being used to manipulate the public opinion, for instance, to justify violence in the present, as it happened in the case of the Russian aggression against Ukraine.

The second justice-related challenge deals with the unequal treatment of certain types of genocide-related information. In some cases, it is attributed to politicisation of archives which can lead to the release of information (e.g. in the form of archival documents) supporting certain political agendas [47] or silencing of information which can be viewed as damaging for a ruling regime [41]. In other cases, the unequal treatment can be related to the belief that some types of information can be less reliable (e.g. due to the assumption that genocide victims can not provide a neutral view on the genocide [21]) or the imbalance between the availability of different types of information (e.g. because of certain groups of individuals being more likely to survive and, thus, leave testimonies [47]).

4 Bridging Memory Ethics and AI-driven IR System Design

After identifying the common ethical challenges of human curation of genocide-related information, it is important to examine to what degree they are applicable for IR system-based curation and how IR system design can be bridged with memory ethics to address concerns associated with these challenges. For this purpose, this section will use the same framework of three groups of challenges related to respect for individual (consent, double vision, and privacy), beneficence (representation and distortion/manipulation), and fairness/justice (politicisation of curation and unequal treatment).

In the case of consent, the difficult part of bridging ethics and IR design relates to consent granting usually being part of the initial stage of data generation (e.g. the recording of a testimony or registering of an account to upload digital materials). One exception here relates to IR systems used in the context of web search, where indexing of digital-born materials is an ongoing process. However, in most cases, the IR systems process data for which consent has already been granted (e.g. in the case of collections stored in the heritage institution or materials generated through the online platform). While it can be possible to integrate consent checks or regular requests for consent re-granting, this specific aspect can arguably be more relevant for the overall model of institution/platform functionality and not necessarily for the IR design.

In terms of double vision, AI-driven IR systems are sometimes argued [30] to be capable of encouraging trust and empathy which can counter the potential dehumanisation of victims associated with this challenge. Such a problem can be particularly pressing with the passing of the living witnesses of the genocides which amplifies the risk of them being increasingly treated as objects and not the subjects of research. One particular example of the use of IR for countering this issue is the shift towards more human-to-human communication-like forms of IR, for instance, the use of conversational agents as a form of curation of genocide-related information.

The potential of IR systems for dealing with privacy-related challenges shares certain similarities with the case of content. In some cases, the matters of privacy are dealt with during the initial state of data generation (e.g. in the case of thorough anonymisation of genocide-related evidence). However, in other cases, IR systems can have a rather ambiguous impact on the privacy of individuals the information about whom they are curating. Advancements in several fields of AI (e.g. computer vision and natural language processing; [63]) enable novel possibilities for recognising the presence of individuals or mentions of specific entities in the data; however, the same advancements can be used for protecting individual privacy (e.g. by masking private information present in the documents). Under these circumstances, the inclusion of particular functionalities in the IR system design can either expose or protect private information. The choice of the functionalities can be informed by examining the selection of materials which the system is expected to work with and its intended uses.

Filtering out privacy-sensitive content can be another alternative to making modifications to the original data. Often associated with the right to be forgotten [58] also known as the right to erasure, this approach can be particularly applicable in the case of IR systems dealing with genocide-related information in the context of web search, where the modification of indexed data is not necessarily possible. Applied for protecting individual privacy this mechanism might be less applicable for genocide-related information, where its evocation for specific cases (e.g. the application of the right to be forgotten for perpetrators of a genocide) might contradict the public interest [59] . However, in other cases (e.g. protecting the victims) it might be essential for tackling privacy-related challenges, thus stressing the importance of functionalities that can facilitate requests for the activation of the right to be forgotten in the IR system design.

From the point of view of beneficence-related challenges, IR can enable new possibilities for addressing the problem of representation. The new formats of curating genocide-related information (e.g. via social bots [30]), as well as more personalised approaches (e.g. the ones taking into consideration the level of knowledge identified on the basis of earlier history of interactions with the IR system), can move the genocide-focused storytelling beyond the traditional modes of representation. While the adequacy of these novel IR approaches for dealing with genocide-related information can be debated, it is important to investigate their potential.

Similar to addressing the problem of representation, AI-driven IR systems can facilitate countering distortion of genocide-related information. The possible approaches for doing it vary from automated detection of distorted information (e.g. the denialist claims) and their subsequent filtering/de-prioritisation (e.g. in the case of Holocaust denial content being countered by commercial platforms) to the provision of contextual information to the system outputs dealing with the genocide. The growing body of research on integrating mechanisms of detecting and countering misinformation in AI-driven IR systems [60, 61] demonstrates possibilities provided by these systems for preventing the distortion of historical facts.

At the same time, it is important to acknowledge the dangers posed by IR systems to the beneficence of curation of genocide-related information, in particular, in the context of the increasing complexity of IR systems [30]. Such complexity makes it more difficult to identify potential instances of system manipulation, in particular, for the users having a limited understanding of the logic behind the system functionality. Together with the limited knowledge about the overall composition of the pool of outputs (e.g. in the case of web search IR systems dealing potentially with billions of genocide-related outputs), it stresses the importance of integrating transparency in the IR system design.

In the case of concerns about archives' politicisation, the impact of AI-driven IR systems can be ambiguous. Depending on how aware about the functionality of IR systems actors involved in politicisation of genocide-related information are and to what degree these actors are capable of influencing the system, IR systems can either facilitate politicisation or counter it. Contextual factors are particularly important for instance, under the condition of intense politicisation of genocide-related information within a particular country, the transparent functionality of IR systems used by the local heritage institutions may actually facilitate the appropriation of the systems for controlling information curation. By contrast, non-transparent IR mechanisms used by a transnational company that is less dependent on the whims of the local memory regime may actually counter politicisation by offering a less politicised selection of information.

The question of the intended use of the IR systems is also of particular importance for identifying their ability to deal with unequal treatment of genocide-related information. Similar to IR systems dealing with news [62], genocide-focused IR systems can serve different normative functions. More deliberative models of IR systems can be optimised for enabling equal representation of genocide-related information (e.g. in terms of visibility of specific aspects of the genocide or particular sites [36]) via either personalised or non-personalised curation, whereas more liberal models might omit the matters of equality, instead giving visibility to a few prominent aspects which the system expects the user to be particularly interested in. The preference for a particular model determines the logic behind the design of a particular IR system; however, determining such a preference might be a rather non-trivial task (e.g. what stakeholder groups shall be able to decide on it?) which is also true for realising more complex models of information curation (e.g. what characteristics to take into consideration when deciding on the equality/lack of equality in representation of specific aspects of a genocide?).

5 Discussion

The chapter scrutinised the ways for bridging AI and memory ethics in the context of IR systems dealing with genocide-related information. Using a Belmont criteria-inspired typology of ethical challenges associated with human curation of information about genocides, it discussed to what degree IR systems can address curation issues related to respect for individuals, beneficence and justice/fairness. The results of this discussion highlight several important points concerning the potential of IR systems for curating information about genocides, both historical (e.g. the Holocaust) and recent ones (e.g. Rohingya genocide).

The first point suggests that the AI-driven IR systems are, unfortunately, not a silver bullet capable of easily solving ethical challenges associated with the curation of genocide-related information. Even while they can address some of the issues related to human curation (e.g. by enabling new possibilities for addressing some of respect-or fairness-related challenges), they can also worsen other issues (e.g. the beneficence-related challenges), in particular, considering the high complexity and frequent lack of transparency of IR systems. Under these circumstances, it becomes of paramount importance to take into consideration the complex relationship between IR systems and memory ethics when designing the former to minimise the possibility of IR having detrimental effects on the lives of individuals affected by genocides and on genocide remembrance.

Second, similar to other domains (e.g. journalism [51]), there is a tradeoff between the realisation of AI potential for enhancing the performance of IR systems (e.g. in terms of addressing ethical challenges, in particular, the ones related to beneficence and justice/fairness) and transparency. While the increased complexity of IR systems enables new possibilities for making the treatment of different groups of genocide victims more fair (e.g. in terms of making their suffering equally visible via AI curation) and dealing with the problem of representation (e.g. in terms of filtering out and removing information distorting historical facts), it also makes the functionality of these systems less transparent, thus limiting the user control over the system [52].

Third, the growing presence of genocide-related information on commercial (and not only heritage-oriented) platforms poses additional difficulties for its curation through IR systems. Because of the generalist focus of commercial platforms (e.g. Google), it is difficult (albeit not impossible as shown by the case of COVID-related information moderation [53]) to enable distinct treatment of specific types of information. Under these circumstances, IR systems used by commercial platforms often treat information about sensitive and traumatic subjects (e.g. genocides) in the same way and following the same logic (e.g. to maximise user engagement as in the case of some social media sites) as other subjects such as entertainment topics. The possibility of such non-differentiated treatment can result in a number of ethics-related issues (in particular, related to the respect for individuals and beneficence of curation) and prompts the importance of the dialogue between the commercial platforms and heritage practitioners as well as other genocide-related actors (e.g. survivors or their families) in order to find a way for addressing these issues.

Finally, it is important to note several limitations of the conducted study. The primary limitation is the reliance on a conceptual approach to discuss the relationship between IR systems and memory ethics. Specifically, the chapter relies on the existing academic scholarship for synthesising the main challenges of human curation of genocide-related information and discussing the possible ways of addressing them through using AI-driven IR systems. Future research can benefit from a more empirically-driven approach (e.g. based on interviews) to solicit opinions of heritage practitioners on the ethics-related issues involved in curation of information about different genocides as well as how these can be affected by IR systems.

A related challenge concerns the focus on the existing research on memory ethics and information curation regarding one particular instance of genocide, namely the Holocaust. While such a focus is not surprising considering the particular importance of the Holocaust, in particular, for the Global North [54], it is crucial to acknowledge that information about other genocides might pose different challenges, in particular, considering the uniqueness of each genocide [55] as well as the increasing criticism of West-oriented standardisation of genocide commemoration [56, 57]. Under these circumstances, it is important not only to extend the discussion of the role of IR systems to other instances of genocide, including the ones occurring in the Global South and Global East, but take into consideration that requirements for AI-driven IR systems in these cases may be different.

References

1. Salton G, McGill, MJ (1983) Introduction to modern information retrieval. Mcgraw-Hill
2. Jones KS (1999) Information retrieval and artificial intelligence. Artif Intell 114(1–2):257–281
3. Boughanem M, Akermi I, Pasi G, Abdulahhad, K (2020) Information retrieval and artificial intelligence. In: A guided tour of artificial intelligence research. Springer, Cham
4. Guo J et al (2020) A deep look into neural ranking models for information retrieval. Inf Process & Manag 57(6)
5. Roetzel PG (2019) Information overload in the information age: a review of the literature from business administration, business psychology, and related disciplines with a bibliometric approach and framework development. Bus Res 12(2):479–522
6. Jones KS (1991) The role of artificial intelligence in information retrieval. J Am Soc Inf Sci 42(8):558–565
7. Hersh WR (2015) Information retrieval for healthcare. In: Reddy C, Aggarwal C (eds) Healthcare data analytics. CRC Press, Boca Raton
8. Miranda A, Miah SJ (2021) Designing an innovative unified contextual architecture for improving information retrieval service in healthcare organizations. Inf Dev. https://doi.org/10.1177/02666669211104949
9. Karimi M, Jannach D, Jugovac M (2018) News recommender systems-Survey and roads ahead. Inf Process & Manag 54(6):1203–1227
10. Mitova E, Blassnig S, Strikovic E, Urman A, Hannak A, de Vreese CH, Esser F (2022) News recommender systems: a programmatic research review. Ann Int Commun Assoc :1–30
11. Alamdari PM, Navimipour NJ, Hosseinzadeh M, Safaei AA, Darwesh A (2020) A systematic study on the recommender systems in the E-commerce. IEEE Access 8:115694–115716
12. Zhang H et al (2020) Towards personalized and semantic retrieval: an end-to-end solution for e-commerce search via embedding learning. In: Proceedings of the 43rd international ACM SIGIR conference on research and development in information retrieval. ACM, New York
13. Schenkolewski-Kroll S, Tractinsky A (2006) Archival description, information retrieval, and the construction of thesauri in Israeli archives. Arch Sci 6(1):69–107
14. Liew CL (2005) Online cultural heritage exhibitions: a survey of information retrieval features. Program 39(1):4–24

15. Jakubowicz A (2009) Remembering and recovering Shanghai: Seven Jewish families [re]-connect in cyberspace. In: Save as… digital memories. Palgrave Macmillan, London
16. Santana Talavera A, Rodríguez Darias AJ, Díaz Rodríguez P, Aguilera Ávila L (2012) Facebook, heritage and tourism reorientation. The cases of Tenerife and Fuerteventura (Canary Isles, Spain). Int J Web Based Communities 8(1):24–39
17. Lechtholz-Zey J (2012) The laws banning Holocaust denial. Genocide Prevention Now 9. Available via Institute on the Holocaust & Genocide in Jerusalem. https://cutt.ly/M1bdkbj. Cited 28 Nov 2022
18. Ware A, Laoutides C (2019) Myanmar's 'Rohingya' conflict: Misconceptions and complexity. Asian Aff 50(1):60–79
19. Makhortykh M (2018) NoKievNazi: social media, historical memory and securitization in the Ukraine crisis. In: Memory and securitization in contemporary Europe. Palgrave Macmillan, London
20. Agarwal K (2016) Doing right online: archivists shape an ethics for the digital age. Available via Perspectives on History. https://cutt.ly/r1bfwQ8. Cited 28 Nov 2022
21. Altanian M (2017) Archives against genocide denialism?. Available via Swisspeace. https://cutt.ly/n1bfxg6. Cited 28 Nov 2022
22. Einwohner RL (2011) Ethical considerations on the use of archived testimonies in Holocaust research: Beyond the IRB exemption. Qual Sociol 34(3):415–430
23. Tesar M (2015) Ethics and truth in archival research. Hist Educ 44(1):101–114
24. Subotić J (2021) Ethics of archival research on political violence. J Peace Res 58(3):342–354
25. Gibson PL, Jones S (2012) Remediation and Remembrance:'Dancing Auschwitz' Collective Memory and New Media. J Commun Stud 5(10):107–131
26. Makhortykh M (2019) Nurturing the pain: audiovisual tributes to the Holocaust on YouTube. Holocaust Stud 25(4):441–466
27. Daelen VV (2019) Data sharing, holocaust documentation and the digital humanities: introducing the European holocaust research infrastructure (EHRI). Umanistica Digitale 3(4):1–9
28. Carter KS, Gondek A, Underwood W, Randby T, Marciano R (2022) Using AI and ML to optimize information discovery in under-utilized, Holocaust-related records. AI & Soc 37:837–858
29. Walden VG (2022) What is 'virtual Holocaust memory'? Mem Stud 15(4):621–633
30. Shur-Ofry M, Pessach G (2019) Robotic collective memory. Wash Univ Law Rev 97:975–1005
31. pravda I (2020) Babyn Yar. The museum of horrors directed by Khrzhanovsky. Available via Istorychna pravda. https://www.istpravda.com.ua/eng/articles/2020/05/14/157507/. Cited 28 Nov 2022
32. Alexander N (2021) Obsolescence, forgotten: "Survivor Holograms", virtual reality, and the future of Holocaust commemoration. Cinergie-Il Cinema e le altre Arti 10(19):57–68
33. Manca S (2021) Digital memory in the post-witness era: how Holocaust museums use social media as new memory ecologies. Inf 12(1):1–17
34. Łysak T (2022) Vlogging Auschwitz: new players in Holocaust commemoration. Holocaust Stud 28(3):377–402
35. Divon T, Ebbrecht-Hartmann T (2022) JewishTikTok: The JewToks' Fight against Antisemitism. In: TikTok cultures in the united states. Routledge, London
36. Makhortykh M, Urman A, Ulloa R (2021) Hey, Google, is it what the Holocaust looked like?. First Monday 26(10). https://doi.org/10.5210/fm.v26i10.11562
37. Wight AC (2020) Visitor perceptions of European Holocaust Heritage: a social media analysis. Tour Manag 81:1–12
38. Kansteiner W (2017) The Holocaust in the 21st century: digital anxiety, transnational cosmopolitanism, and never again genocide without memory. In: Digital memory studies. Routledge, London
39. Manca S, Passarelli M, Rehm M (2022) Exploring tensions in Holocaust museums' modes of commemoration and interaction on social media. Technol Soc 68:1–13
40. Guhl J, Davey J (2020) Hosting the 'Holohoax': a snapshot of Holocaust denial across social media. Available via The Institute for Strategic Dialogue. https://cutt.ly/G1bxtsR. Cited 28 Nov 2022

41. Khumalo NB (2019) Silenced genocide voices in Zimbabwe's archives: drawing lessons from Rwanda's post-genocide archives and documentation initiatives. Inf Dev 35(5):795–805
42. Hawkes M (2012) Containing testimony: Archiving loss after Genocide. Contin 26(6):935–945
43. Anabo IF, Elexpuru-Albizuri I, Villardón-Gallego L (2019) Revisiting the Belmont Report's ethical principles in internet-mediated research: perspectives from disciplinary associations in the social sciences. Ethics Inf Technol 21(2):137–149
44. Fujii LA (2010) Shades of truth and lies: interpreting testimonies of war and violence. J Peace Res 47(2):231–241
45. Hirsch M (2001) Surviving images: Holocaust photographs and the work of postmemory. Yale J Crit 14(1):5–37
46. Jacobs JL (2004) Women, genocide, and memory: the ethics of feminist ethnography in Holocaust research. Gend & Soc 18(2):223–238
47. Luft A (2020) How do you repair a broken world? Conflict (ing) archives after the Holocaust. Qual Sociol 43(3):317–343
48. Crane SA (2008) Choosing not to look: representation, repatriation, and holocaust atrocity photography. Hist Theory 47(3):309–330
49. Lang B (2000) Holocaust representation: art within the limits of history and ethics. JHU Press, Baltimore
50. Atkins SE (2009) Holocaust denial as an international movement. ABC-CLIO, Santa Barbara
51. Bastian M, Makhortykh M, Dobber T (2019) News personalization for peace: how algorithmic recommendations can impact conflict coverage. Int J Confl Manag 30(3):309–328
52. Storms E, Alvarado O, Monteiro-Krebs L (2022) 'Transparency is Meant for Control' and Vice Versa: learning from Co-designing and Evaluating Algorithmic News Recommenders. Proc ACM Hum Comput Interact 6(CSCW2):1–24
53. OECD (2020) Combatting COVID-19 disinformation on online platforms. Available via OECD. https://cutt.ly/O1bYUGJ. Cited 28 Nov 2022
54. Assmann A (2010) The Holocaust—A global memory? Extensions and limits of a new memory community. In Memory in a global age. Palgrave Macmillan, London
55. Jonassohn K (1998) Genocide and gross human rights violations: in comparative perspective. Transaction Publishers, Piscataway
56. David L (2017) Against standardization of memory. Hum Rights Q 39(2):296–318
57. Dubey I (2021) Remembering, forgetting and memorialising: 1947, 1971 and the state of memory studies in South Asia. India Rev 20(5):510–539
58. Esposito E (2017) Algorithmic memory and the right to be forgotten on the web. Big Data & Soc 4(1):1–11
59. Makhortykh M (2021) Memoriae ex machina: how algorithms make us remember and forget. Georg J Int Aff 22(2):180–185
60. Ozbay FA, Alatas B (2020) Fake news detection within online social media using supervised artificial intelligence algorithms. Phys Stat Mech Appl 540:1–19
61. Fernández-Pichel M, Losada DE, Pichel JC (2022) A multistage retrieval system for health-related misinformation detection. Eng Appl Artif Intell 115:105211
62. Helberger N (2019) On the democratic role of news recommenders. Digit J 7(8):993–1012
63. Curzon J, Kosa TA, Akalu R, El-Khatib K (2021) Privacy and artificial intelligence. IEEE Trans Artif Intell 2(2):96–108

Algorithmic Fairness in Multi-stakeholder Platforms

Gourab K. Patro

Abstract Major online platforms today are multi-stakeholder in nature and they cater to the interests of various stakeholders. Examples include e-commerce platforms with sellers of goods and services and buyers who pay for them, hotel booking platforms with hosts and guests, media-streaming platforms artists or content creators and viewers, and many more. The focus here is on the information access services (like search and recommendation) deployed on these platforms. While traditionally these services were designed in consumer-centric ways, they are found to be unfair to providers. Since the providers depend on these platforms for their livelihood, fairness for providers is an important and necessary design element in many scenarios. This chapter summarizes the domain and discusses various prior works on provider fairness in such multi-stakeholder platforms after a brief review of major works on algorithmic fairness in machine learning. Many recent works show that provider fairness can cause loss of utility and unfairness for consumers. Following this, a number of works have proposed fairness notions for both providers and consumers in different settings, and studied multi-stakeholder fairness to balance various fairness and utility goals or constraints. This chapter reviews some major works on multi-stakeholder fairness in the following three aspects: the problem setting, fairness notions, and proposed approaches. It also discusses how most of the works on multi-stakeholder fairness have considered settings with only two stakeholders, gives some examples on other platforms with more than two stakeholders, and how they are fundamentally different in terms of the utility structure for the stakeholders.

Keywords Algorithmic fairness · Multi-stakeholder platforms · Multi-stakeholder fairness

---

G. K. Patro (✉)
L3S Research Center, Leibniz University Hannover and Indian Institute of Technology Kharagpur, Kharagpur, India
e-mail: patrogourab@gmail.com

A. Mukherjee et al. (eds.), *Ethics in Artificial Intelligence: Bias, Fairness and Beyond*, Studies in Computational Intelligence 1123,
https://doi.org/10.1007/978-981-99-7184-8_5

1 Introduction

Major online platforms today are multi-stakeholder in nature and they cater to the interests of various stakeholders. For example, e-commerce platforms with sellers of goods and services and buyers who pay for them; hotel booking platforms with hosts and guests; media-streaming platform artists or content creators and viewers; ride-hailing platforms with ride providers and interested riders; hiring and gig-economy platforms with employers and candidates or workers; and food delivery and other hyperlocal platforms with vendors, buyers, and delivery agents. These platforms often have two or more major stakeholders and the information access services like search, recommendation, and matching help them interact [46, 64]. This chapter will focus more on search and recommendation services on such platforms with two stakeholders: providers and consumers.

Traditionally the search and recommendation services were designed in consumer-centric ways, trying to optimize for consumer satisfaction [16–18]. However, recently such approaches are found to be unfair to providers leading to significant discrepancies in provider utilities [10, 19, 20, 22, 26]. Since the providers often depend on these platforms for their livelihood, fairness for providers has become an important and necessary design element in many scenarios. Some works have studied this problem of provider fairness and proposed various approaches [30, 34, 38, 39] (more details in Sect. 3). However, recent works [46, 51, 58] have found that making the system fair for providers can cause unfairness for consumers (inequalities in consumer utilities) along with significant losses in consumer utilities, and vice versa (more details in Sect. 4). Following this, a number of works [46, 50–65] have proposed fairness notions for both providers and consumers in different settings, and studied multi-stakeholder fairness to balance various fairness and utility goals or constraints (more details in Sect. 5).

This chapter has been written as an introductory review of the algorithmic fairness in machine learning, online platforms, and major works on fairness in multi-stakeholder platforms and multi-stakeholder fairness. The chapter is organized as follows. Section 2 begins the discussion by reviewing various works on general algorithmic fairness, fairness in machine learning, and briefly discusses its connection with fairness in online platforms. Then, Sect. 3 follows up and talks about fairness in online platforms. However, for multi-stakeholder platforms, multiple fairness notions and utility metrics have to be used, making the task more complicated (more in Sect. 4). Section 4 also lists the major fairness and utility metrics, and the conflicts and agreements among them. It also discusses how most of the works on multi-stakeholder fairness have considered settings with only two stakeholders, gives some examples of other platforms with more than two stakeholders, and how they are fundamentally different in terms of the utility structure for the stakeholders. Finally, Sect. 5 reviews the general multi-stakeholder fairness problem in the following three aspects: the problem setting, fairness notions, and proposed approaches.

2 Algorithmic Fairness and Its Importance

Today various automated data-driven decision-making systems (especially machine learning systems) are increasingly used in many real-world applications including criminal justice [1], banking [2], hiring, and admissions [3, 4]. The performance and behavior of these systems usually have significant implications on people's lives thereby largely shaping our world. The algorithms used in these systems are often vulnerable to various biases existing in their training data or specifics sometimes overlooked during the design of the algorithms [5–7]. Due to this they can show various kinds of unfair behavior: for example, an AI-based criminal risk assessment system, COMPAS, falsely flagged black defendants as future criminals, wrongly labeling them at almost twice the recidivism risk as white defendants, and mislabeled white defendants as low risk more often than black defendants [6]; various facial recognition softwares have shown disproportionately low accuracy for darker skinned and Asian faces [8, 9]; Facebook's ad delivery system disparately delivers STEM career ads to different gender groups [10]. Such disparate behavior often has significant adverse impact on people's lives and may to lead to further disparity in society. Following such cases with unfairness, bringing algorithmic fairness in such systems has now become an important goal, especially when designing algorithmic systems to be deployed at scale.

A range of studies have looked into how to avoid undesired disparate behavior in various algorithmic systems and ensure fairness. In machine learning context (also extensible to other settings), various definitions of fairness have been proposed. Most of them can be put into two main categories: individual and group fairness (detailed surveys: [11, 12]).

- **Group fairness** notions [13, 14] try to ensure similar effect on different demographic groups (often legally protected classes like gender and race).
- **Individual fairness** notions [15] try to ensure that similar individuals receive similar outcomes.

Note that in regular machine learning systems which decide on whether a defendant gets bail [6], whether an applicant gets loan [2], etc., the individuals (defendants or applicants) have a clear benefit from getting a positive decision in their favor. Similarly, in many online platforms, algorithmic systems perform tasks like recommendation and ranking of items or content, ad allocations, etc. which also translate to beneficial outcomes for different stakeholders (e.g., sellers get more sales by getting their items ranked on top). Thus while the main motivation for fairness in online platforms has been similar to that in regular machine learning setup, i.e., to ensure fair outcomes for the stakeholders, there is also an added motivation from a resource allocation perspective which will be discussed in the next section.

3 Algorithmic Fairness in Online Platforms

Major online platforms today include e-commerce platforms like Amazon, Ebay, and Flipkart; media-streaming platforms like YouTube and Spotify; ride-hailing platforms like Uber, Grab, and Ola; hotel booking platforms like Booking.com and Airbnb; hiring and gig-economy platforms like LinkedIn and Freelancer; food delivery platforms like Foodpanda, Grubhub, and Zomato; and many more. These platforms often have a set of providers of items or services which are then consumed by consumers. The platforms themselves act as mediators between providers and consumers. Essentially, the consumers use various information retrieval services like search, recommendation, and advertisements on the platform to find relevant items or services. In fact, ranking and recommendation systems are ubiquitous across such platforms. These systems are traditionally designed in a consumer-centric way, i.e., trying to optimize consumer satisfaction or utility [16–18]. However, recent studies [19, 20] have found that such designs can exacerbate selection bias [21–23] and result in phenomena like popularity bias [20, 24, 25] where a few providers (usually old and big or a certain privileged group of providers) receive most of the consumer attention while the remaining (usually small and new or less privileged providers) providers struggle to get consumer attention. Other such undesirable effects include racial bias in Airbnb [26], gender-based discrimination in the delivery of STEM career ads [10], racial and gender bias in freelance marketplaces [19], and disproportionately high visibility of old and popular providers [27].

Note that the consumer attention on these platforms gives exposure to providers, and it often translates to different socio-economic opportunity and livelihood (e.g., sales on e-commerce, jobs on gig-economy platforms) for the providers [28, 29]. In addition, the total available exposure on a platform is often a limited resource (based on the total anticipated consumer traffic). Thus, any bias in the performance of the information retrieval services leads to bias in the distribution of exposure (also called as *exposure bias* [30, 31]), thereby being unfair to the providers and undermining their well-being. In fact any bias in the core relevance scoring model gets exacerbated by the presence of position bias [32] (the phenomenon of higher consumer attention to top ranked items) in online platforms [33, 34]. Therefore, there is a need to ensure that these retrieval services are fair to the providers while still ensuring good consumer satisfaction. Following this, a range of studies have tried to develop mechanisms for provider fairness in ranking and recommendation (see [22, 33, 35–37] for surveys).

While the central motivation here has been to provide the providers a fair access to opportunity on the platform, the mechanisms used for provider fairness in the recent works can be categorized into two major branches based on their approach [33].

- **Proportionality in top-*k***: A set of works [38–40] have proposed mechanisms to ensure probability-based fairness by ensuring certain proportional (e.g., an equal proportion or a proportion based on an actual population distribution) placement of items or individuals from a protected group in the top-*k* ranking positions in a search or recommendation result. These works have focused more on group fair-

ness for providers (e.g., ensuring fair distribution of candidates in top-*k* results provided to an employer—here a consumer). Here, group fairness simply corresponds to fair representation from the protected groups.
- **Fairness of exposure**: Another set of works [30, 34, 41] formalize the exposure metric (the total attention received from the consumers) for providers by assigning numerical values to each ranking or recommendation positions based on expected attention or click probability. Then, for provider fairness, these works propose mechanisms (both deterministic and stochastic) to ensure fair distribution of exposure either among individuals or groups of providers. Fairness of exposure can be used to ensure exposure of individual providers remains proportional (i.e., individual fairness) to their relevance scores or to ensure exposure of groups of individuals remains proportional (i.e., group fairness) to a desired fair proportion like their cumulative relevance score and their distribution in the population.

While the above-mentioned works clearly establish the need for provider fairness in online platforms, there is more to fairness in multi-stakeholder online platforms than just provider fairness and consumer satisfaction. We discuss this in the next section.

4 The Case of Multi-stakeholder Platforms

This section sheds light on metrics (other than just provider fairness and consumer satisfaction) which are important for design and analysis of various multi-stakeholder platforms and how their interplay makes algorithmic fairness on such platforms a hard task. Note that while online platforms like e-commerce, media streaming, ride-hailing, and hotel booking platforms in the last section often have only two types of stakeholders (providers and consumers), various gig-economy, food delivery, and other hyperlocal platforms have three or more types of stakeholders (e.g., Uber Eats has restaurants or providers, delivery agents, and consumers as three types of stakeholders). Next, Sects. 4.1 and 4.2 discuss about platforms with only two types of stakeholders and platforms with more than two types of stakeholders, respectively.

4.1 Platforms with Two Types of Stakeholders

The focus, here, is on the platforms with two kinds of stakeholders, the providers who list their item or services on the platform and the consumers who consume them or pay for them. Examples of such platforms include e-commerce, media-streaming platforms, ride-hailing, and hotel booking platforms. Since the item corpus on these platforms is often huge, the search and recommendation services are most of times the only way consumers explore the platform and find relevant items. Thus one can say that these services match providers to consumers through item or content retrieval. Figure 1 depicts such platforms as two-sided platforms. Here we shall focus

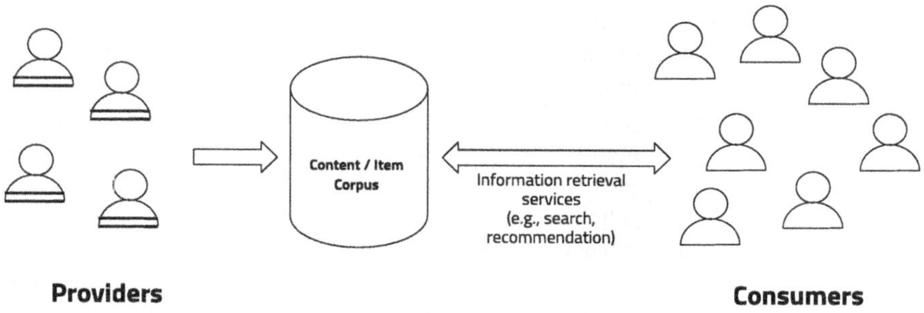

Fig. 1 Two-sided platforms with two types of stakeholders: providers and consumers

Table 1 Examples of two-sided online platforms

Platform	Item/content consumed	Providers	Consumers
E-commerce (Amazon, Ebay)	Items for sale	Sellers	Buyers
Hiring (LinkedIn, Naukri)	Candidate profiles	Candidates	Employers
Job search (LinkedIn, Naukri)	Job ads	Employers	Candidates
Media streaming (Spotify, Tidal)	Media content	Artists	Listeners
Hotel booking (Booking.com, Airbnb)	Rooms	Owners or operators	Tenants

on such two-sided platforms where the two sides (stakeholders) benefit from successful matching done through the platform's retrieval services. Note that depending on the core item or service being consumed on a platform, the providers and consumers can be identified (see Table 1 for example).

4.1.1 Relevant Fairness and Utility Metrics

In the two-sided platforms with providers and consumers as stakeholders, there are four major metrics for fairness and utility as listed below:

- **Consumer utility**: Retrieval systems often use various statistical or machine learned models to estimate the relevance of the items to consumer's search or recommendation query [42, 43]. Traditionally these services are designed to optimize for consumer utility, and select or rank top-k items based on relevance following the *probability ranking principle* [16]. Metrics like mean reciprocal rank and discounted cumulative gain [44, 45] also follow the same principle. Note that

the relevance scoring models are often evaluated through user studies and A/B testing.
- **Provider utility**: Providers on two-sided platforms gain from getting consumer attention and thereby platform exposure for their items or content. While many works have considered simple exposure as a form of utility for the providers [30, 31, 34, 46], others argue that only the exposure from possibly interested consumers' attention should count toward provider utility via careful estimation of position bias [32, 33].
- **Provider fairness**: For provider fairness, various forms of parity can be considered as fairness depending on the context and application domain. For example, in settings like marketplaces, individual providers need to be ensured exposure based on their relevance [30], this can be characterized as an extension of individual fairness [15]. On the other hand, in settings like hiring and gig-economy platforms, workers or candidates from different sensitive groups need to get equal or proportional exposure [34, 39, 47]—a group fairness notion.
- **Consumer fairness**: While consumer utility metrics often represent their satisfaction, the relevance scoring models behind the metrics might have different performance levels for different consumers or consumer groups. This disparity may arise due to biases in data [48]. Thus, consumer fairness notions are introduced to reduce the inequality in consumer satisfaction [48, 49]. For example, in case of hiring or admission platforms, disparity in the quality of job or college listings being served to different groups of candidates (like high-paying jobs being shown only to a certain group of candidates) can create societal disparity in the long run and is therefore not at all desirable [48].

4.1.2 Observed Conflicts and Agreements

The utility and fairness metrics often have conflicts and agreements among each other. Various works have found trade-offs between these metrics. Some are discussed next.

- **Consumer utility versus provider utility**: When relevant items or services are served to right consumers, it often proves to be beneficial for both corresponding providers and consumers. Thus, ideally, it should equally improve both consumer and provider utility [33]. However, in reality, true relevance is unknown, and it is estimated using various relevance scoring mechanisms. Thus, these relevance scores become core of consumer utility metrics. On the other hand, apart from relevance, the position of an item in ranked results also affects the amount of consumer attention received by the item (position bias [32]), thereby also the consumer click probability. Thus, position bias affects both consumer and provider utility. However for simplicity, many works [30, 34, 41] consider only the click probability of a provider's items (only based on its position in the results) as its utility.
- **Consumer utility versus provider fairness**: This is one of the first conflicts observed on various platforms which essentially gave rise to various works (as

discussed in Sect. 3) on provider fairness in online platforms. Various works in different settings [10, 19, 20, 30, 38] show that optimizing for consumer utility can undermine provider fairness. In fact, biases in models for consumer utility optimization can cause unfairness issues on the provider side, leading to popularity bias [24, 25], polarization [49], racial and gender discrimination [19]. On the other hand, a retrieval system, looking to ensure provider fairness, is often found to be losing average consumer utility [30, 34, 46, 50].

- **Provider fairness versus consumer fairness**: Recent works on two-sided platforms [46, 51–54] show that there are conflicts between provider-side and consumer-side fairness. It has been observed that since provider fairness tries to reduce the inequality on the provider side, it causes losses in consumer utilities. However, these losses may not always be equal for all consumers which then causes inequality and unfairness on the consumer side. Similarly, when consumer fairness is maintained (e.g., equity of consumer utilities), provider fairness would get impacted [46, 51]. A trivial case would be to use the consumer-centric approach to retrieval systems which often causes disparity on provider side as discussed in Sect. 3.
- **Consumer fairness versus consumer utility**: A few works have studied only consumer-side fairness [48, 49]. Consumer fairness notions defined based on inequality in underestimation and overestimation (of relevance scores) errors are found to improve overall consumer utility since they reduce underestimation and overestimation errors [48]. Similar evidence is found in other works too [49].

4.2 Platforms with Three or More Types of Stakeholders

While most of the works on multi-stakeholder platforms have focused more on two-stakeholder (especially two-sided) platforms, some platforms do have more than two types of stakeholders. Examples include food delivery and other hyperlocal platforms like Uber eats, JustEat, and Deliveroo which have three kinds of stakeholders: vendors who offer food and other items, buyers who order the items, and delivery agents who facilitate the delivery of orders. Similarly media-streaming platforms like YouTube have creators of video content, viewers who consume the content, and advertisers who may or may not directly contribute to platform's content creation, but are essentially the main source of revenue for the creators. This kind of setting where the source of socio-economic benefits of providers (creators) comes from a third kind of stakeholders instead of directly coming from content consumption is what distinguishes platforms like YouTube from platforms like Spotify. Thus, the provider utility metric can significantly differ on YouTube and Spotify. Apart from the fairness and utility metrics for the providers and consumers, a whole new set of fairness and utility metrics for the third kind of stakeholders would need to be considered. While the three-stakeholder setting itself is quite complicated to be formally modeled, further addition of stakeholders would of course increase the complexity of the problem. Thus, there has been limited research work on fairness in such settings,[1] but of course be taken up as significant future works.

5 Multi-stakeholder Fairness

Following the conflicts among consumer utility, provider fairness and consumer fairness on two-sided platforms, many recent works have studied and tried to design ranking and recommendation mechanisms to improve performance on all the metrics [46, 50–65]. Table 2 lists the major research works (in chronological order) on two-sided fairness along with their problem settings, fairness notions, and proposed approaches. These listed works are summarized below.[2]

Problem settings: The problem settings of the works on two-sided fairness can be categorized into *personalized versus non-personalized, offline versus online or dynamic*, and *ranked list versus unranked list versus matching*. Most of the works have looked into the regular personalized ranking, recommendation, or matching where the two-sided fairness serves the two stakeholders (providers and consumers). One recent work [54] has utilized the two-sided fairness (in non-personalized recommendation) simply as two fairness notions, one serving the consumers and another for ideological balancing of the recommendation, serving an interest of the platform itself. While personalized ranking or recommendation is available on most of the platforms today, some non-personalized recommendations (all consumers get same set of recommendations) are still popular in certain platforms (e.g., top news on news platforms, top trending topics on Twitter, and top selling mobile phones). Many works [46, 51, 52, 56, 60–63] have studied ranking and recommendation in an offline setting where either a set of consumers are given or a distribution of consumer query intents is given, and the proposed method decides the ranking or recommendation for all consumers or consumer queries at the same time, not worrying about the sequence of these queries. Such offline recommendations can be directly used in mass recommendations through email or app notifications or in platforms which update ranking/recommendations periodically instead of dynamically, or simply be extended to online dynamic settings where the set of active consumers can be selected at a point of time, and recommendations can be given. Such works have given the foundational concepts and set the agenda for two-sided fairness. A few works [50, 53, 58, 59] have extended the ideas to online dynamic setting as well. Many earlier works [46, 53, 54, 56] have considered an unranked list creation task (especially for

[1] Abdollahpouri and Burke [55] briefly talk about one of such settings and consider the intermediary stakeholders as side stakeholders.

[2] Note that since most of the relevant research works have focused on two-sided platforms, here we limit our discussion to only two-sided platforms while studies on various three-sided platforms (as discussed in Sect. 4.2) can be a significant future research agenda.

Table 2 Research works on two-sided fairness

Paper	Provider fairness	Consumer fairness	Problem setting	Approach
Balanced neighborhoods 2018 [56]	Parity in outcomes of provider **groups** (item neighborhood balancing for representative collaborative filtering)	Parity in utilities of consumer **groups** (consumer neighborhood balancing for representative collaborative filtering)	Offline recommendation (each side's fairness studied separately, not together)	Sparse linear method for representative collaborative influence
FairRec 2020 [46], FairRec+ 2021 [57]	**Individual** exposure-based fairness (minimum exposure guarantee)	Parity of losses in **individual** consumer utilities (utility envyfreeness)	Offline recommendation	Unique greedy fair allocation mechanism
Basu et al. 2020 [58]	**Group** fairness of exposure	Parity in utilities of consumer **groups**	Online ranking on marketplaces	Linear programming
Incremental fairness 2020 [53]	Gradual changes in **individual** provider exposure	Minimum utility guarantee all **individual** consumers	Platform recommendation system upgrades	Integer programming
Fairness in repeated matchings 2020 [59]	Fair distribution of income to **individual** drivers (income equality or income proportional to activity)	Fair distribution of resultant delays of waiting times among **individual** riders	Matching market (ride-hailing)	Integer linear programming
Safe and sustainable recommendation 2020 [50]	**Individual** exposure-based fairness (Minimum exposure guarantee)	Minimum utility guarantee to **individuals**	Online local recommendations	Linear programming-based matching
Wang et al. 2021 [60]	Fair (proportional to overall merit) distribution of **group** exposures	Parity in **group** fairness	Offline ranking	Convex optimization of a stochastic ranking policy followed by Birkhoff–von Neumann
Mondal et al. 2021 [54]	Fair representation of political or ideological (**groups**) representation in top-k	Fair representation of (**individual**) consumers' choices in top-k	Top-k non-personalized recommendation (trending news, #tags)	Fair voting
Lorenz dominant ranking 2021 [51]	Increase in the exposure of the worst-off item (**individual**)	Increase in the utility of the worst-off item **individual**	Offline item recommendation and reciprocal recommendation (ranked)	Convex welfare function optimization
Tfrom 2021 [52]	Exposure-based fairness (**individual**)	Parity of losses in **individual** consumer utilities	Offline and online ranking and recommendation	Greedy re-ranking
CPFair 2022 [61, 62]	Exposure-based fairness (**individual**)	Parity in utilities of **individual** consumers	Offline ranked recommendation	Mixed-integer linear programming
Wu et al. 2022 [63]	Exposure-based fairness (popularity and genre **groups**)	Parity in utilities of consumer **groups**	Offline ranked recommendation	Multiple gradient descent for multi-objective optimization

recommendation) where the effect of position of an item in a recommended list does not cause any change in provider or consumer utilities. Subsequent works [50–52, 58, 60, 61, 63] have extended them to ranked list creation (for search and recommendation). Another recent work [59] has studied two-sided fairness in a matching setting (especially for ride-hailing platforms).

Fairness notions: Above-mentioned works on two-sided fairness have used various fairness notions taking inspiration from earlier works on fairness in machine learning (as detailed in Sect. 2) and also classical theories of justice and fairness in economics and related studies. Majority of fairness notions utilize the (in)equality in utilities of a particular side or stakeholder. The area of fairness in machine learning has also inspired the research to define fairness notions for individuals or groups on each stakeholder's side. Depending on the specific settings, the research has followed either individual or group fairness notions. For example, in e-commerce marketplace setting, providers and consumers are usually treated in individual manner, thus individual notions of fairness make more sense [46]. On the other hand, in platforms like charity project recommendations (e.g., Kiva.org), popularity bias may cause huge disparity by significantly overfunding projects from developed world and underfunding projects from underdeveloped world; thus a group fairness notion for the providers (charity organizations managing the projects) would be desired [56]. The type of fairness notions (individual or group) of each research work has been highlighted in bold text in Table 2. Moreover, most of the works have considered simple parity-based notions for stakeholder individuals or groups—extending the concepts from machine learning fairness research—a few works have taken inspiration from other fields, e.g., minimum exposure guarantee for providers [46, 50] from welfare economics, maxi-min share for providers and envyfreeness for consumers [46] from fair allocation, incrementalism from moral philosophy [53], fair representation from voting theory [54], Lorenz efficiency for provider and consumer fairness [51] from economics.

Approaches: With regard to approaches, many of the works have proposed the following types of approaches for the two-sided fairness problem in their chosen settings: greedy re-ranking [46, 52], linear programming-based matching [50, 59], fair voting [54], linear programming [58], integer programming [56, 61, 62], convex optimization [51, 60], and gradient descent-based optimization [63].

6 Conclusion

This chapter started with general introduction to algorithmic fairness, its need in various algorithmic settings, followed by a focused discussion on algorithmic fairness in online platforms. Next parts provided detailed discussions on multi-stakeholder platforms, fairness and utilities of various stakeholders, their conflicts and agreements, and finally multi-stakeholder fairness. Since the domain of fairness in multi-stakeholder platforms is still in early stage, this chapter can serve as an introductory review of relevant works which have motivated, inspired, or lead to the works on fairness in multi-stakeholder platforms, and the need, models, and approaches for multi-stakeholder fairness. While most of the relevant works are on platforms with just two stakeholders (i.e., two-sided fairness), extending them to fairness in various platforms with three or more types of stakeholders can be a significant future contribution.

References

1. Rigano C (2019) Using artificial intelligence to address criminal justice needs. Natl Inst Justice J 280:1–10
2. Mukerjee A, Biswas R, Deb K, Mathur AP (2002) Multi-objective evolutionary algorithms for the risk-return trade-off in bank loan management. Int Trans Oper Res 9(5):583–597
3. Cohen L, Lipton ZC, Mansour Y (2020) Efficient candidate screening under multiple tests and implications for fairness. In: 1st symposium on foundations of responsible computing
4. Marcinkowski F, Kieslich K, Starke C, Lünich M (2020) Implications of AI (un-) fairness in higher education admissions: the effects of perceived AI (un-) fairness on exit, voice and organizational reputation. In: Proceedings of the 2020 conference on fairness, accountability, and transparency, pp 122–130
5. Ntoutsi E, Fafalios P, Gadiraju U, Iosifidis V, Nejdl W, Vidal M-E, Ruggieri S, Turini F, Papadopoulos S, Krasanakis E et al (2020) Bias in data-driven artificial intelligence systems-an introductory survey. Wiley Interdiscip Rev Data Min Knowl Discov 10(3):1356
6. Angwin J, Larson J, Mattu S, Kirchner L (2016) Machine bias
7. O'Neil C (2017) Weapons of math destruction: how big data increases inequality and threatens democracy (Crown publishers, New York, 2016). Coll Res Libr 78(3):403–404
8. Racial discrimination in face recognition technology. https://sitn.hms.harvard.edu/flash/2020/racial-discrimination-in-face-recognition-technology/. Accessed 01 Dec 2022
9. Are face-detection cameras racist?. https://content.time.com/time/business/article/0,8599,1954643,00.html. Accessed 01 Dec 2022
10. Lambrecht A, Tucker C (2019) Algorithmic bias? an empirical study of apparent gender-based discrimination in the display of stem career ads. Manag Sci 65(7):2966–2981
11. Mehrabi N, Morstatter F, Saxena N, Lerman K, Galstyan A (2021) A survey on bias and fairness in machine learning. ACM Comput Surv (CSUR) 54(6):1–35
12. Finocchiaro J, Maio R, Monachou F, Patro GK, Raghavan M, Stoica A-A, Tsirtsis S (2021) Bridging machine learning and mechanism design towards algorithmic fairness. In: Proceedings of the 2021 ACM conference on fairness, accountability, and transparency, pp 489–503
13. Corbett-Davies S, Pierson E, Feller A, Goel S, Huq A (2017) Algorithmic decision making and the cost of fairness. In: Proceedings of the 23rd ACM SIGKDD international conference on knowledge discovery and data mining, pp 797–806
14. Hardt M, Price E, Srebro N (2016) Equality of opportunity in supervised learning. Adv Neural Inf Process Syst 29
15. Dwork C, Hardt M, Pitassi T, Reingold O, Zemel R (2012) Fairness through awareness. In: Proceedings of the 3rd innovations in theoretical computer science conference, pp 214–226
16. Robertson SE (1977) The probability ranking principle in IR. J Doc
17. Bobadilla J, Ortega F, Hernando A, Gutiérrez A (2013) Recommender systems survey. Knowl Based Syst 46:109–132
18. Schütze H, Manning CD, Raghavan P (2008) Introduction to information retrieval, vol 39
19. Hannák A, Wagner C, Garcia D, Mislove A, Strohmaier M, Wilson C (2017) Bias in online freelance marketplaces: Evidence from taskrabbit and fiverr. In: Proceedings of the 2017 ACM conference on computer supported cooperative work and social computing, pp 1914–1933

20. Bellogín A, Castells P, Cantador I (2017) Statistical biases in information retrieval metrics for recommender systems. Inf Retr J 20(6):606–634
21. Heckman J (1990) Varieties of selection bias. Am Econ Rev 80(2):313–318
22. Chen J, Dong H, Wang X, Feng F, Wang M, He X (2020) Bias and debias in recommender system: a survey and future directions. arXiv:2010.03240
23. Baeza-Yates R (2018) Bias on the web. Commun ACM 61(6):54–61
24. Abdollahpouri H (2019) Popularity bias in ranking and recommendation. In: Proceedings of the 2019 AAAI/ACM conference on AI, ethics, and society, pp 529–530
25. Ahanger AB, Aalam SW, Bhat MR, Assad A (2022) Popularity bias in recommender systems-a review. In: International conference on emerging technologies in computer engineering. Springer, pp 431–444
26. Edelman B, Luca M, Svirsky D (2017) Racial discrimination in the sharing economy: evidence from a field experiment. Am Econ J Appl Econ 9(2):1–22
27. Salganik MJ, Dodds PS, Watts DJ (2006) Experimental study of inequality and unpredictability in an artificial cultural market. Sci 311(5762):854–856
28. Graham M, Hjorth I, Lehdonvirta V (2017) Digital labour and development: impacts of global digital labour platforms and the gig economy on worker livelihoods. Transf Eur Rev Labour Res 23(2):135–162
29. Hesmondhalgh D, Osborne R, Sun H, Barr K (2021) Music creators' earnings in the digital era. Intellectual property office research paper forthcoming
30. Biega AJ, Gummadi KP, Weikum G (2018) Equity of attention: amortizing individual fairness in rankings. In: The 41st international ACM SIGIR conference on research & development in information retrieval, pp 405–414
31. Banerjee A, Patro GK, Dietz LW, Chakraborty A (2020) Analyzing 'near me'services: Potential for exposure bias in location-based retrieval. In: 2020 IEEE international conference on big data (big data). IEEE, pp 3642–3651
32. Agarwal A, Zaitsev I, Wang X, Li C, Najork M, Joachims T (2019) Estimating position bias without intrusive interventions. In: Proceedings of the twelfth ACM international conference on web search and data mining, pp 474–482
33. Patro GK, Porcaro L, Mitchell L, Zhang Q, Zehlike M, Garg N (2022) Fair ranking: a critical review, challenges, and future directions. In: 2022 ACM conference on fairness, accountability, and transparency, FAccT '22. Association for Computing Machinery, New York, NY, USA, pp 1929–1942
34. Singh A, Joachims T (2018) Fairness of exposure in rankings. In: Proceedings of the 24th ACM SIGKDD international conference on knowledge discovery & data mining, pp 2219–2228
35. Wang Y, Ma W, Zhang* M, Liu Y, Ma S (2022) A survey on the fairness of recommender systems. ACM Trans Inf Syst
36. Ekstrand MD, Das A, Burke R, Diaz F (2022) Fairness in recommender systems, pp 679–707
37. Zehlike M, Yang K, Stoyanovich J (2021) Fairness in ranking: a survey. arXiv:2103.14000
38. Zehlike M, Bonchi F, Castillo C, Hajian S, Megahed M, Baeza-Yates R (2017) Fa* ir: a fair top-k ranking algorithm. In: Proceedings of the 2017 ACM on conference on information and knowledge management, pp 1569–1578
39. Celis LE, Straszak D, Vishnoi NK (2018) Ranking with fairness constraints. In: ICALP
40. Asudeh A, Jagadish H, Stoyanovich J, Das G (2019) Designing fair ranking schemes. In: Proceedings of the 2019 international conference on management of data, pp 1259–1276
41. Zehlike M, Castillo C (2020) Reducing disparate exposure in ranking: a learning to rank approach. In: Proceedings of the web conference 2020, pp 2849–2855
42. Liu T-Y et al (2009) Learning to rank for information retrieval. Foundations and Trends® in Information Retrieval 3(3):225–331
43. Adomavicius G, Tuzhilin A (2005) Toward the next generation of recommender systems: a survey of the state-of-the-art and possible extensions. IEEE Trans Knowl Data Eng 17(6):734–749
44. Voorhees EM et al (1999) The trec-8 question answering track report. In: Trec, vol 99, pp 77–82

45. Järvelin K, Kekäläinen J (2002) Cumulated gain-based evaluation of IR techniques. ACM Trans Inf Syst (TOIS) 20(4):422–446
46. Patro GK, Biswas A, Ganguly N, Gummadi KP, Chakraborty A (2020) Fairrec: two-sided fairness for personalized recommendations in two-sided platforms. In: Proceedings of the web conference 2020, pp 1194–1204
47. Geyik SC, Ambler S, Kenthapadi K (2019) Fairness-aware ranking in search & recommendation systems with application to linkedin talent search. In: Proceedings of the 25th ACM SIGKDD international conference on knowledge discovery & data mining, pp 2221–2231
48. Yao S, Huang B (2017) Beyond parity: fairness objectives for collaborative filtering. Adv Neural Inf Process Syst 30
49. Rastegarpanah B, Gummadi KP, Crovella M (2019) Fighting fire with fire: using antidote data to improve polarization and fairness of recommender systems. In: Proceedings of the twelfth ACM international conference on web search and data mining, pp 231–239
50. Patro GK, Chakraborty A, Banerjee A, Ganguly N (2020) Towards safety and sustainability: designing local recommendations for post-pandemic world. In: Fourteenth ACM conference on recommender systems, pp 358–367
51. Do V, Corbett-Davies S, Atif J, Usunier N (2021) Two-sided fairness in rankings via Lorenz dominance. Adv Neural Inf Process Syst 34:8596–8608
52. Wu Y, Cao J, Xu G, Tan Y (2021) Tfrom: A two-sided fairness-aware recommendation model for both customers and providers. In: Proceedings of the 44th international ACM SIGIR conference on research and development in information retrieval, pp 1013–1022
53. Patro GK, Chakraborty A, Ganguly N, Gummadi K (2020) Incremental fairness in two-sided market platforms: on smoothly updating recommendations. In: Proceedings of the AAAI conference on artificial intelligence, vol 34, pp 181–188
54. Mondal AS, Bal R, Sinha S, Patro GK (2021) Two-sided fairness in non-personalised recommendations (student abstract). In: Proceedings of the AAAI conference on artificial intelligence, vol. 35, pp 15851–15852
55. Abdollahpouri H, Burke R (2019) Multi-stakeholder recommendation and its connection to multi-sided fairness. arXiv:1907.13158
56. Burke R, Sonboli N, Ordonez-Gauger A (2018) Balanced neighborhoods for multi-sided fairness in recommendation. In: Conference on fairness, accountability and transparency. PMLR, pp 202–214
57. Biswas A, Patro GK, Ganguly N, Gummadi KP, Chakraborty A (2021) Toward fair recommendation in two-sided platforms. ACM Trans Web (TWEB) 16(2):1–34
58. Basu K, DiCiccio C, Logan H, Karoui NE (2020) A framework for fairness in two-sided marketplaces. arXiv:2006.12756
59. Sühr T, Biega AJ, Zehlike M, Gummadi KP, Chakraborty A (2019) Two-sided fairness for repeated matchings in two-sided markets: a case study of a ride-hailing platform. In: Proceedings of the 25th ACM SIGKDD international conference on knowledge discovery & data mining, pp 3082–3092
60. Wang L, Joachims T (2021) User fairness, item fairness, and diversity for rankings in two-sided markets. In: Proceedings of the 2021 ACM SIGIR international conference on theory of information retrieval, pp 23–41
61. Naghiaei M, Rahmani HA, Deldjoo Y (2022) Cpfair: personalized consumer and producer fairness re-ranking for recommender systems. In: SIGIR '22, New York, NY, USA
62. Naghiaei M, Rahmani HA, Deldjoo Y (2022) PyCPFair: a framework for consumer and producer fairness in recommender systems. Softw Impacts 13:100382
63. Wu H, Ma C, Mitra B, Diaz F, Liu X (2022) A multi-objective optimization framework for multi-stakeholder fairness-aware recommendation. ACM Trans Inf Syst (TOIS)
64. Burke R (2017) Multisided fairness for recommendation. arXiv:1707.00093
65. Sonboli N, Burke R, Ekstrand M, Mehrotra R (2022) The multisided complexity of fairness in recommender systems. AI Mag 43(2):164–176

Biases and Ethical Considerations for Machine Learning Pipelines in the Computational Social Sciences

Suparna De, Shalini Jangra, Vibhor Agarwal, Jon Johnson, and Nishanth Sastry

Abstract Computational analyses driven by Artificial Intelligence (AI)/Machine Learning (ML) methods to generate patterns and inferences from big datasets in computational social science (CSS) studies can suffer from biases during the data construction, collection and analysis phases as well as encounter challenges of generalizability and ethics. Given the interdisciplinary nature of CSS, many factors such as the need for a comprehensive understanding of different facets such as the policy and rights landscape, the fast-evolving AI/ML paradigms and dataset-specific pitfalls influence the possibility of biases being introduced. This chapter identifies challenges faced by researchers in the CSS field and presents a taxonomy of biases that may arise in AI/ML approaches. The taxonomy mirrors the various stages of common AI/ML pipelines: dataset construction and collection, data analysis and evaluation. By detecting and mitigating bias in AI, an active area of research, this chapter seeks to highlight practices for incorporating responsible research and innovation into CSS practices.

S. De (✉) · S. Jangra · V. Agarwal · N. Sastry
University of Surrey, Guildford, UK
e-mail: s.de@surrey.ac.uk

S. Jangra
e-mail: s.jangra@surrey.ac.uk

V. Agarwal
e-mail: v.agarwal@surrey.ac.uk

N. Sastry
e-mail: n.sastry@surrey.ac.uk

J. Johnson
University College, London, UCL, UK
e-mail: jon.johnson@ucl.ac.uk

A. Mukherjee et al. (eds.), *Ethics in Artificial Intelligence: Bias, Fairness and Beyond*, Studies in Computational Intelligence 1123,
https://doi.org/10.1007/978-981-99-7184-8_6

1 Introduction

Advances in communication networks and the growing use of social networking platforms mean that there is an unprecedented amount of information that provides an important source for understanding a population [1, 2]. Computational tools have been successfully used to analyse the resulting structured and unstructured data, with the aim of understanding individuals, groups and their social practices. This well-studied field of computational social science (CSS) is characterized by (1) the involvement of human subjects, with the resulting capabilities and tools also impacting individuals and communities, (2) the use of large and complex datasets, drawn from mixed methods data collection, incorporating both self-reporting through surveys and experiments, as well as through observation of 'unconstrained' behaviour on social media platforms and (3) application of AI or ML-driven computational or algorithmic solutions to the resulting big data to generate insights, inferences and predictions about human behaviours, social networks and systems.

We cannot use ML predictive models in a black-box fashion for social science problems [24]. It is necessary to analyse the ethical implications and consequences of these models' output as these may have real-world consequences and impacts. Due to this human impact, computational research needs to be "ethical, trustworthy and responsible" [3]. However, this very human nature of the data means that it encounters issues of representativeness, uniformity and bias [1]. Thus, this chapter focuses on some of the key issues around ethics and generalizability confronting CSS researchers in the age of big data. These issues are analysed through the lens of the data lifecycle in ML pipelines, as identified in existing literature [4], i.e. covering dataset creation/collection, data analysis and data (model) evaluation, as shown in Fig. 1. This is followed by a discussion of the strategies and existing initiatives to address the issue of bias in CSS ML pipelines.

2 Dataset Creation and Collection Bias

The first stage of a typical ML pipeline starts with data collection, which can take the form of scraping it from social networking platforms, e.g. Reddit [38, 39] and Kialo [36, 37] or creating a dataset from available survey data collection APIs [5]. The creation and archiving of such complex datasets naturally give rise to issues of data privacy and de-identification, necessitating steps for individual privacy protection and conforming to laws and principles of informed consent (e.g. GDPR[1]).

The following sub-sections describe how biases can be introduced in the ML pipeline during the dataset creation and collection phase, which also includes labelling or annotating the data.

[1] https://eugdpr.org/.

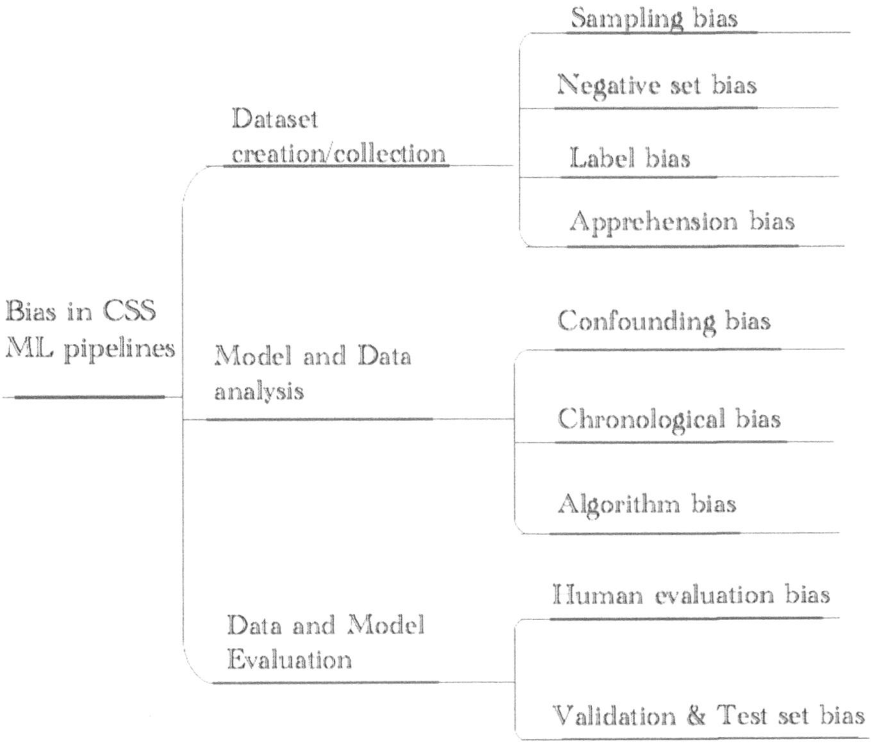

Fig. 1 Taxonomy of biases in CSS ML pipelines

2.1 Sampling Bias

One of the most common instances of dataset bias is sampling bias, which occurs due to some types of instances being selected more than others [4]. Datasets are often created with a particular set of instances, with most social media research using a sample of all available data to make inferences about a larger population [47]. With the sampling methods necessitating representativeness of both demographics and behaviour, any systemic distortion in the sampled data, due to sparsity for instance, can compromise its representativeness. It is also difficult to obtain a uniform random sampling from social platforms. Sparsity in the data can also be magnified due to platform characteristics, for instance, by limiting the length of users' posts which in turn affect data retrieval [50]. Therefore, poor generalization of the trained AI models can be an unintended consequence of sampling bias.

The probability of gaps in data coverage also increases in the case of longitudinal studies spanning decades, as in the case of the UK's MRC National Survey of Health and Development (MRC/NSHD) study, which has a lower occurrence of labelled instances in some vocabulary categories such as measures of psychological well-being, omics and sleep, compared to more recent birth cohort studies [6].

This sampling bias can also occur due to missing instances or features in the datasets and socio-cultural conditions of data generation. The importance of context around the social and historical conditions in the data generation process is also crucial where observational data may have 'non-random missingness' [3] and meaningful noise.

2.2 Negative Set Bias

Negative set bias occurs when there are not enough samples representative of the remaining world (negative instances which are not present in the dataset). The ever-increasing use of social platforms and behaviour capture by both private corporations and government bodies has led to unprecedented amounts of data being collected on human activities' traces. However, historically disadvantaged groups are often "less involved in the formal economy and its data generating activities" [8], which means that there are not enough samples representing such groups in the dataset, causing negative set bias. This leads to potential reinforcement of digital divides and data inequities through biased techniques that render digitally marginalized groups invisible.

Negative set bias may also be manifested due to user *self-selection* bias, either due to users exercising self-censorship [51, 53], e.g. not 'liking' or sharing/deleting a post despite reading it, due to privacy concerns. It can also occur due to platform characteristics which makes some user activities invisible, e.g. dataset only includes users who post content, not those who only read it [52].

A related ethical question is that most of the data harvesting occurs without the conscious "consent or active awareness of the people whose digital and digitalised lives are the targets of surveillance, consumer curation, and behavioural steering" [3], raising questions of privacy, autonomy and meaningful consent. An example can be found in the geo-tagging capabilities of some social networking platforms, with some users unaware of their posts being geo-tagged, while others consciously using geo-tagging to 'advertise' where they have been [54]. Negative set bias also has real-world implications when the resultant analyses are used to inform data-driven public policies, which may be geared towards economically advantaged and data-rich areas [55].

The opposite of this 'negative' bias is in domains such as hate speech, where there are insufficient positive samples (most datasets have very few hate speech occurrences). This was a problem for us in the work with MPs [70] and also more recently in the Decentralized Web [71]. Vidgen et al. [72] have a unique approach to this problem—they artificially generate (through crowd workers) a balanced dataset on hate speech.

2.3 Label Bias

Label bias is bias associated with the labelling or data annotation process. Subjective biases and domain background of the annotators can deeply influence the annotation process, leading to inconsistencies in the labelling process. Different annotators have different perspectives based on their different life experiences and world view [40]. For example, annotating hate speech is a highly subjective task [41]. Often, different annotators give different labels to instances based on their varying levels of sensitivity towards a particular hate type. Their aggregated labels, mostly using majority voting, are often treated as gold labels in various hate speech datasets and therefore, favour majority opinions [42]. ML models trained over these datasets with label biases can be highly biased in nature and can result in poor performance in detecting hate speech accurately. Guest et al. [69] replace majority voting with facilitated meetings between annotators to improve the quality of the datasets generated. AnnoBERT [73] directly incorporates subjectivity into a hate speech detection model and shows that this improves classification performance.

The subjectivity of the labelling process also contributes to its propensity towards bias, which can be magnified in the case of high-volume longitudinal studies, as reported in our recent work [6], as the labels given for an object type can diverge significantly more than where the data collection period is short. As reported in this work, unsupervised topic modelling approaches uncovered instances of unintuitive manual labelling in cases of semantic overlaps in question texts, with the mislabelled instances reflecting the domain background of the human labellers.

2.4 Apprehension Bias

Apprehension bias is concerned with how user behaviour (and hence, how it is manifested in the resulting dataset) is impacted by the awareness of being observed. In response to observers such as other platform users or administrators, users may choose different behaviours of *self-presentation*, which is termed as online 'Hawthorne effect' [47]. Such effects have been studied in location-based social networks, where check-ins at public locations such as restaurants are more likely than at private ones such as a doctor's surgery [61]. Conditioning of individual writing style of reviews has been found to be influenced by prior ratings and reviews [62].

Apprehension bias is also prominent in observational CSS, where study participants are recruited for administering surveys or questionnaires, as this brings into play the researchers as active observers, which can cause a behavioural change as a conscious response to being studied. This is illustrated in the mixed-mode data collection stage for the National Child Development Study survey, as reported in [7]. The authors of this work report not only variance in participation rates between telephone-only and Web-based respondents to the survey, but also differences in response values which can be attributable to the mode of data collection. For instance, Web-based participants had a higher non-response to questions related to finances, and also had more negative stances to self-rated subjective parts of the study, such as health and well-being. As a result, the authors identified the potential for subsequent biases in the analyses, and recommended techniques to correct for these.

3 ML Model and Data Analysis Bias

A second realm of problems concerns the construction of the algorithms (if they are structured and not completely self-learning), and the selection of features or criteria. Biases can be introduced through untrue assumptions of the distribution of the data, the data cleaning and pre-processing methods as well as the choice of the ML models.

3.1 Confounding Bias

Confounders are external variables that manipulate the estimate of the apparent relationship between the independent variable of interest and the dependent (output) variable and hence lead to erroneous output of the model [26]. A confounding variable can influence the outcome of an experiment in various ways, such as invalid correlations, increasing variance and suggesting an association where none exists or masking a true association. Confounding, sometimes referred to as confounding bias, is mostly described as a "mixing or blurring of effects" [27]. For instance, [32] states that the root reason for the bias in recommender systems present in e-commerce (e.g. Amazon and Alibaba) websites and social networking platforms such as Twitter or Facebook are confounder variables that influence both which items the user will interact with and how they rate them. Approaches to address the detrimental effects of confounding variables include those by Liu et al. [33] who proposed a debiased information bottleneck (DIB) objective function to reduce the confounding bias in the biased feedback without having to retrain with unbiased data. Randomization such as random initialization or random choices during learning is the only way to control for confounding because it will balance measured and unmeasured confounding.

A type of confounding bias is that of 'omitted variable', where the analysis is carried out without considering the relevant features. This is more significant for predictive ML, such as regression analysis, when the omitted variables match the independent variables or regressors and the dependent variables are determined by this omitted variable [43]. This causes the analysis to correlate their effects to model variables that caused bias, to the estimated effects, thus, confounding the cause-effect relationship, making it challenging to differentiate between "attributes that merely correlate and those that are causally related" [47]. An example is the spurious correlation between URLs in tweets and their retweet rates, which were found to be due to the URLs often co-occurring with hashtags [65]. Consequences of omitted variable bias include both exaggerating and underrating the effect in the analysis, flipping the statistical analysis result or even causing an effect to be hidden in the outcome.

A related concept to omitted variable bias is 'proxy' or indirect bias, with variables used as proxies for sensitive ones, or those that are not directly measurable. The use of proxy variables abounds in CSS analyses, though they may suffer from validity or reliability issues [47]. For social networks research, interest in a topic is often indirectly measured through the proxy variable of number of posts on the topic [46], though it fails to conclusively capture how much content of the topic is actually read. The choice of proxy participants to determine user traits or demographic criteria has also been shown to influence the performance of prediction models, for example, in the case of using university alumni registered on a social platform as proxy for 'young' college graduates to determine their views on a new law [48], which resulted in an important source of bias.

3.2 Chronological Bias

Chronological bias refers to the change in study design that happens over time and affects the study results, due to temporal variations caused by population drifts or system drifts [56].

System drifts can lead to issues of 'temporal validity' of the study conclusions as illustrated in the case of the Google Flu Trends (GFT) platform, which, following an algorithm update in 2009, made headlines in 2013 for predicting more than double the number of doctor visits for flu-like illness versus that reported by the Centers for Disease Control and Prevention (CDC). An analysis into the GFT over-estimation [49] revealed issues with the algorithm dynamics and changes made in the underlying Google search algorithm in June 2011 and February 2012. The analysis uncovered that Google's modifications in search results in 2011/12, to suggest additional search terms and also potential diagnoses for searches, tracked closely with GFT errors when comparing correlated search terms for the GFT time series to those returned by the CDC data.

Population drifts occur when study participants whose data is mined or analysed earlier during an intervention are subject to different social exposures or are at a different risk from participants who are recruited later [9]. This has been exemplified with studies on both Facebook [57] and Twitter [58] social platforms. Changes in platform users' lifestyles and evolution of online communities [59] can also affect how long users are engaged with a topic, which may also be dependent on changes in the platform itself, such as the addition of new features.

3.3 Algorithm Bias

Algorithm bias is defined as bias that is solely induced or added by the algorithm; for instance, a ML model that relies on randomness for fair distributions of results is not truly random.

Specific types of such bias include *ranking* bias—privileging some algorithmic results more than others in the way they are presented. For instance, social media platforms employ algorithms designed to promote trending content that may negatively affect the overall quality of information on the platform. As an extension to ranking bias, personalization algorithms employed in social media platforms and search engines are designed to select only the most engaging and relevant content for each individual user. But in doing so, it "may end up reinforcing the cognitive and social biases of users" [60], with less diverse exposure to content, thus making them part of a social bubble and more vulnerable to manipulation.

Another case is of *insensitive measure* bias [9] that can result from the use of an insufficiently accurate method to detect the outcome of interest, where the method is not sensitive enough to detect true differences. Examples include use of automated Natural Language Processing (NLP) tools for dependency parsing and language detection, which may not be robust when different dialects, which vary from the mainstream languages, are present in the dataset [64]. The use of alternative objective functions, when the true criterion is not directly measurable, such as user clicks as a substitute for user satisfaction [66], has the potential of creating 'Matthew effects' of self-reinforcing feedback loops [8] between datasets, decisions and algorithms. Such effects can have harmful downstream consequences such as false negatives disappearing from the dataset [8], with the resulting asymmetry skewing the decision-making process.

Social media platforms also expose users to a less diverse content from a significantly narrower spectrum of sources compared to non-social media sites like Wikipedia [44]. This is called *homogeneity* bias. This can take the form of 'gatekeeping', where there is a distinct preference for some topics, and 'coverage', with differences in attention given to certain topics as well as how these are presented [63]. Pre-trained language representations such as Bidirectional Encoder Representations from Transformers (BERT), which is trained on a general-purpose corpus, may under- or over-represent the relationship between different words in the dataset under analysis, as even though different scientific domains may use the same language, the words may have very different semantic connotations.

4 Data and Model Evaluation Bias

4.1 Human Evaluation Bias

Biases during evaluation can be introduced by circumstances of *confirmation bias* (interpreting information that is consistent with existing beliefs), *peak end effect* (cognitive bias related to how subjects remember, by focusing the recall on the 'peak' or more intense moments) and prior beliefs (e.g. culture). For instance, the detailed advertising tools built into many social networking platforms play into the hands of actors looking to spread disinformation by tailoring messages to people who are already inclined to believe them, i.e. exploiting confirmation bias [45]. Human evaluators are also limited by how accurately or by how much information they can recollect, which can result in *recall bias*. People show this bias when they reminiscence information selectively (by omitting details), or when they understand/assess it in a biased way. Schwind et al. [28] state that the pattern of results observed in selection behaviour is also apparent in evaluation behaviour. This implies that the reduction of evaluation bias will occur only when preference-inconsistent recommendations are combined with low prior knowledge conditions.

4.2 Validation and Test Set Bias

Validation and test set bias refer to systematically under- or over-estimating the predictive performance of the model [68]. Practitioners introduce bias into their model when tuning new models based on the performance of old models on the test/holdout data. For instance, developers make changes to the model based on what they have learned about how previous topologies and hyperparameters affected the model accuracy on the test data, thereby introducing bias into the model. By leveraging observations gained from the model's performance on test data, it is possible to optimize the model using the whole dataset, avoiding ever training the model straight on the test samples. The presence of bias in models can be influenced by the samples and labels chosen in the validation and test datasets [4]. Evaluation bias can also arise from inadequate benchmarks or datasets used for testing. Consequently, metrics computed over the whole test or validation set may not always provide a correct indication of the model's fairness.

5 Responsible Research for CSS ML Pipelines

Several initiatives exist, such as "Datasheets for Datasets" [10] for documenting essential information about datasets for training ML models, as part of a move towards practical guidelines for reducing potential bias in AI systems. Others aimed

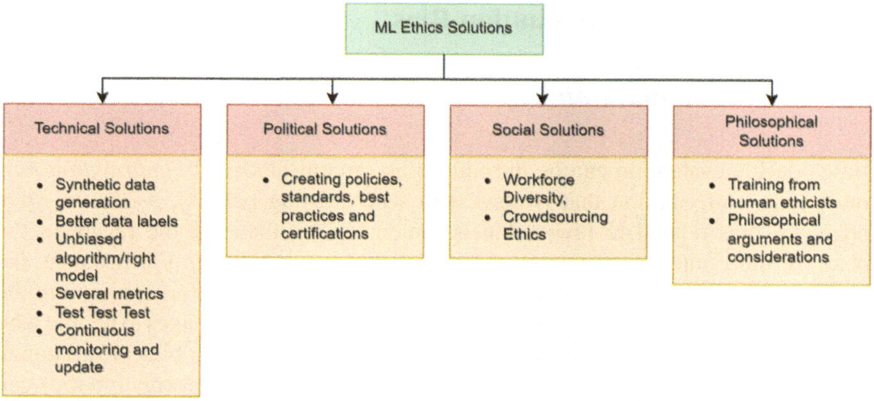

Fig. 2 Ethical solutions to reduce Bias in Machine Learning

at a technical level include initiatives such as "discrimination-aware data mining" (DADM) [8].

Strategies for choosing ML models that may be less discriminatory than baseline choices can include adversarial debiasing [11], where the model learns to predict the outcomes to prevent another adversary AI model from guessing the protected variables based on the outcomes. Another strategy is the dynamic upsampling of training data [12], with the data from underrepresented groups being given more weight during the training phase.

Approaches for reducing CSS bias can be divided into three categories: (i) pre-processing approaches [14], (ii) in-processing approaches [15, 16] and (iii) post-processing approaches [17]. Pre-processing approaches target the foremost source of bias, i.e. data. Their prime objective is to generate a balanced and fair dataset that results in less discriminative ML models [13]. These approaches include altering the data distribution by sampling, re-weighting or modifying the individual training instance. Modelling classification problems with fairness constraints [18], restricting the learner's behaviour by enforcing independence on sensitive features [19] and adversarial debiasing [11] are some of the different in-processing bias-mitigation approaches. Post-processing approaches are applied once the model has been trained on the data, which includes changing the model's internals (white-box approaches) [22, 23] or its predictions (black-box approaches) [17, 20, 21]. Bias-mitigation approaches should provide the middle ground between the ML model's accuracy and fairness.

The solutions proposed for ethical approaches to reduce CSS biases are of four types: technical, social, political and philosophical [29], as depicted in Fig. 2.

1. **Technical Solutions**: One of the prime technical solutions for mitigating bias is synthetic data generation. Generating synthetic data involves defining and setting the parameters of a fair dataset and then generating data that fulfils that definition, which may help protect people's sensitive information. Mislabelling due

to preconceived notions and assumptions of labellers can have unintentional or detrimental real-world consequences. More nuanced labels or categories can help introduce fairness in the system. Additional contextual metadata, for example, the characteristics of the population and the mode of data collection, may also be used to identify and mitigate potentially unmeasured biases. Users' configuration of the algorithm could reflect their cultural and experiential biases. Therefore, having absolute transparency on how the algorithm works can be helpful in designing unbiased algorithms. Training, followed by masking, emphasizes helping marginalized groups while treating the non-sensitive features as the same as others [31]. Setting up the correct parameters, regular spot checks and continuous testing and monitoring also help.

2. **Political Solutions**: It is necessary to establish political control over the ethical deployment of AI/ML to reduce the likelihood of unintended consequences. This can be achieved by creating guidelines, policies and legal frameworks and introducing certifications to learn best practices for the responsible use of these technologies. For instance, the EU passed the General Data Protection Regulation (GDPR) in April 2016 which came into effect in 2018. It mandates organizations to provide citizens in the EU with the "right to explanation", which refers to the right to receive an explanation for an algorithm's output. The government's investment in ethical ML technology research can help to build a knowledgeable workforce capable of developing and deploying ethical machine learning systems.
3. **Social Solutions**: Raising awareness among the public can be an approach towards tackling ML bias. For instance, Google and Microsoft researchers founded the workshop "Fairness, Accountability, and Transparency in Machine Learning" (FAT ML) to examine the repercussions of algorithmic bias. It has now developed into the ACM Conference on Fairness, Accountability, and Transparency (ACM FAccT),[2] which brings together researchers and practitioners from across computer science, law, social sciences and humanities to tackle issues in this area. The involvement of people from diverse populations in the entire ML pipeline will also reduce discrimination. Radford et al. [67] argue that biases emerge throughout the entire ML pipeline that cannot be remedied solely through technical solutions. They describe the way social theory, for example, critical race theory and feminist theory, can help in removing ML bias by providing a framework for understanding the social and cultural contexts in which the data is produced and used.
4. **Philosophical Solutions**: Considering all contextual divergences, humans are the final piece to making ethical decisions. Although machine learning may produce practical and advanced applications, more is needed to replace the human capacity for domain expertise. Gnjatović et al. [34] focused on reintroducing humans into the learning loop. For instance, false, inaccurate or incomplete information floated online in news, social media and on the Web causes societal harm. Reference [35] discusses the importance of hybrid approaches to fighting against online misinformation and disinformation. Both ML tools and humans—including specialized

[2] https://facctconference.org.

professionals and lay persons sourced through crowdsourcing platforms—should collaborate to mitigate the issue.

Acknowledgements This research is funded by the UKRI Strategic Priority Fund as part of the wider Protecting Citizens Online programme (Grant number: EP/W032473/1) associated with the National Research Centre on Privacy, Harm Reduction and Adversarial Influence Online (REPHRAIN), and by the Science and Technology Facilities Council (STFC) DiRAC-funded "Understanding the multiple dimensions of prediction of concepts in social and biomedical science questionnaires" project, grant number ST/S003916/1.

References

1. Shah DV, Cappella JN, Neuman WR (2015) Big data, digital media, and computational social science: possibilities and perils. Ann Am Acad Politic Soc Sci 659(1):6–13. https://doi.org/10.1177/0002716215572084
2. De S, Jassat U, Grace A, Wang W, Moessner K (2022) Mining composite spatio-temporal lifestyle patterns from geotagged social data. In: IEEE international conferences on internet of things (iThings) and IEEE green computing & communications (GreenCom) and IEEE cyber, physical & social computing (CPSCom) and IEEE smart data (SmartData) and IEEE congress on cybermatics (Cybermatics). Espoo, Finland, pp 444–451
3. Leslie D (2022) Don't "research fast and break things": on the ethics of computational social science. arXiv, abs/2206.06370
4. Ramya Srinivasan R, Chander A (2021) Biases in AI systems: a survey for practitioners. ACM Queue 19(2)
5. De S, Moss H, Johnson J, Li J, Pereira H, Jabbari S (2022) Engineering a machine learning pipeline for automating metadata extraction from longitudinal survey questionnaires. IASSIST Quart 46(1)
6. Sharifian-Attar De S, Jabbari S, Li J, Moss H, Johnson J (2022) Analysing longitudinal social science questionnaires: topic modelling with BERT-based embeddings. In: Proceedings of 2022 ieee international conference on big data, Osaka, Japan, 2022, pp 5558–5567. https://doi.org/10.1109/BigData55660.2022.10020678
7. Goodman A, Brown M, Silverwood RJ, Sakshaug JW, Calderwood L, Williams J, Ploubidis George B (2022) The impact of using the Web in a mixed-mode follow-up of a longitudinal birth cohort study: evidence from the national child development study. J Roy Stat Soc: Ser A (Stat Soc) 185(3):822–850
8. Herzog L (2021) Algorithmic bias and access to opportunities. In: Véliz C (ed) The oxford handbook of digital ethics. https://doi.org/10.1093/oxfordhb/9780198857815.013.21
9. Spencer EA, Heneghan C (2017) Catalogue of bias collaboration. In: Catalogue of bias. https://catalogofbias.org/biases/
10. Gebru T, Morgenstern J, Vecchione B, Wortman Vaughan J, Wallach H, Daumé III H, Crawford K (2021) Datasheets for datasets. Commun ACM 64(12):86–92. https://doi.org/10.1145/3458723
11. Zhang BH, Lemoine B, Mitchell M (2018) mitigating unwanted biases with adversarial learning. In: Artificial intelligence, ethics, and society conference
12. Cofone IN (2019) Algorithmic discrimination is an information problem. Hastings Law J 70:1389–1444
13. Ntoutsi E, Fafalios P, Gadiraju U, Iosifidis V, Nejdl W, Vidal ME, ... Staab S (2020) Bias in data-driven artificial intelligence systems-an introductory survey. Wiley Interdiscip Rev: Data Min Knowl Discov 10(3): e1356
14. Hajian S (2013) Simultaneous discrimination prevention and privacy protection in data publishing and mining. arXiv:1306.6805

15. Fish B, Kun J, Lelkes ÁD (2016) A confidence-based approach for balancing fairness and accuracy. In Proceedings of the 2016 SIAM international conference on data mining. Society for Industrial and Applied, pp 144–152
16. Kamishima T, Akaho S, Sakuma J (2021) Fairness-aware learning through regularization approach. In: 2011 IEEE 11th international conference on data mining workshops. IEEE, pp 643–650
17. Hardt M, Price E, Srebro N (2016) Equality of opportunity in supervised learning. Adv Neural Inf Process Syst 29
18. Celis LE, Huang L, Keswani V, Vishnoi NK (2019) Classification with fairness constraints: a meta-algorithm with provable guarantees. In: Proceedings of the conference on fairness, accountability, and transparency, pp 319–328
19. Kamishima T, Akaho S, Asoh H, Sakuma J (2012) Fairness-aware classifier with prejudice remover regularizer. In: Joint European conference on machine learning and knowledge discovery in databases. Springer, Berlin, Heidelberg, pp 35–50
20. Agarwal A, Beygelzimer A, Dudík M, Langford J, Wallach H (2018) A reductions approach to fair classification. In: International conference on machine learning. PMLR, pp 60–69
21. Canetti R, Cohen A, Dikkala N, Ramnarayan G, Scheffler S, Smith A (2019) From soft classifiers to hard decisions: how fair can we be?. In: Proceedings of the conference on fairness, accountability, and transparency, pp 309–318
22. Pedreschi D, Ruggieri S, Turini F (2009) Measuring discrimination in socially-sensitive decision records. In: Proceedings of the 2009 SIAM international conference on data mining. Society for Industrial and Applied Mathematics, pp 581–592
23. Calders T, Kamiran F, Pechenizkiy M (2009) Building classifiers with independency constraints. In: 2009 IEEE international conference on data mining workshops, pp 13–18
24. Wallach H (2018) Computational social science $\neq$ computer science + social data. Commun ACM 61(3):42–44
25. Garcia M (2017) Racist in the machine: the disturbing implications of algorithmic bias. World Policy J 33(4):111–117
26. Zhao Q, Adeli E, Pohl KM (2020) Training confounder-free deep learning models for medical applications. Nat Commun 11(1):1–9
27. Jager KJ, Zoccali C, Macleod A, Dekker FW (2008) Confounding: what it is and how to deal with it. Kidney Int 73(3):256–260
28. Schwind C, Buder J (2012) Reducing confirmation bias and evaluation bias: when are preference-inconsistent recommendations effective-and when not?. Comput Hum Behav 28(6):280–2290
29. Shadowen N (2019) Ethics and bias in machine learning: a technical study of what makes us "good". The transhumanism handbook. Springer, Cham, pp 247–261
30. Shankar S, Halpern Y, Breck E, Atwood J, Wilson J, Sculley D (2017) No classification without representation: Assessing geodiversity issues in open data sets for the developing world. arXiv:1711.08536
31. Ghili S, Kazemi E, Karbasi A (2019) Eliminating latent discrimination: train then mask. Proc AAAI Conf Artif Intell 33(01): 3672–3680
32. He M, Hu X, Li C, Chen X, Wang J (2022) Mitigating confounding bias for recommendation via counterfactual inference. In: Proceedings of the European conference on machine learning and principles and practice of knowledge discovery in databases (ECML-PKDD22)
33. Liu D, Cheng P, Zhu H, Dong Z, He X, Pan W, Ming Z (2021) Mitigating confounding bias in recommendation via information bottleneck. In: Fifteenth ACM conference on recommender systems, pp 351–360
34. Gnjatović M, Maček N, Adamović S (2020) Putting humans back in the loop: a study in human-machine cooperative learning. Acta Polytech Hungarica 17(2)
35. Demartini G, Mizzaro S, Spina D (2020) Human-in-the-loop artificial intelligence for fighting online misinformation: challenges and opportunities. IEEE Data Eng Bull 43(3):65–74
36. Agarwal V, Joglekar S, Young AP, Sastry N (2022) GraphNLI: a graph-based natural language inference model for polarity prediction in online debates. In: Proceedings of the ACM web conference 2022, pp 2729–2737

37. Young AP, Joglekar S, Agarwal V, Sastry N (2022) Modelling online debates with argumentation theory. ACM SIGWEB newsletter, (Spring), pp 1–9
38. Agarwal V, Young AP, Joglekar S, Sastry N (2022) A graph-based context-aware model to understand online conversations. arxiv:2211.09207
39. Guest E, Vidgen B, Mittos A, Sastry N, Tyson G, Margetts H (2021) An expert annotated dataset for the detection of online misogyny. In: Proceedings of the 16th conference of the European chapter of the association for computational linguistics: main volume, pp 1336–1350
40. Akhtar S, Basile V, Patti V (2020) Modeling annotator perspective and polarized opinions to improve hate speech detection. In: Proceedings of the AAAI conference on human computation and crowdsourcing, pp 151–154
41. Aroyo L, Dixon L, Thain N, Redfield O, Rosen R (2019) Crowdsourcing subjective tasks: the case study of understanding toxicity in online discussions. In: Companion proceedings of the 2019 World Wide Web conference, pp 1100–1105
42. Sheng VS, Zhang J, Gu B, Wu X (2017) Majority voting and pairing with multiple noisy labeling. IEEE Trans Knowl Data Eng 1355–1368
43. Wilms R, Mäthner E, Winnen L, Lanwehr R (2021) Omitted variable bias: a threat to estimating causal relationships. Methods Psychol 5:2021
44. Nikolov D, Oliveira DF, Flammini A, Menczer F (2015) Measuring online social bubbles. Peer J Comput Sci 1:e38
45. Ciampaglia GL, Menczer F (2018) Misinformation and biases infect social media, both intentionally and accidentally. The Conversation, 20
46. Chen J, Nairn R, Nelson L, Bernstein M, Chi E (2010) Short and tweet: experiments on recommending content from information streams. In: Proceedings of the SIGCHI conference on human factors in computing systems (CHI '10), New York, NY, USA, pp 1185–1194
47. Olteanu A, Castillo C, Diaz F, Kıcıman E (2019) Social data: biases, methodological pitfalls, and ethical boundaries. Front Big Data 2
48. Cohen R, Ruths D (2013) Classifying political orientation on twitter: It's not easy!. Proc Int AAAI Conf Web Soc Media 7(1):91–99
49. Lazer D, Kennedy R, King G, Vespignani A (2014) The parable of google flu: traps in big data analysis. Science 343(6176):1203–1205
50. Naveed N, Gottron T, Kunegis J, Alhadi AC (2011) Searching microblogs: coping with sparsity and document quality. In: Proceedings of the 20th ACM international conference on information and knowledge management, CIKM '11, New York, pp 183–188
51. Gong W, Lim E-P, Zhu F, Cher PH (2016) On unravelling opinions of issue specific-silent users in social media. In: Proceedings of the international AAAI conference on web and social media, Cologne
52. Das S, Kramer A (2013) Self-censorship on facebook. In: Proceedings of the international AAAI conference on web and social media, Boston, MA
53. Wang Y, Norcie G, Komanduri S, Acquisti A, Leon PG, Cranor LF (2011) 'i regretted the minute i pressed share': a qualitative study of regrets on facebook. In: Proceedings of the seventh symposium on usable privacy and security, SOUPS '11, New York, NY, pp 10:1–10:16
54. Tasse D, Liu Z, Sciuto A, Hong J (2017) State of the geotags: motivations and recent changes. In: Proceedings of the international AAAI conference on web and social media, Montreal, QC
55. Hecht B, Stephens M (2014) A tale of cities: urban biases in volunteered geographic information. In: Proceedings of the international AAAI conference on web and social media, Ann Arbor, M
56. Salganik MJ (2017) Bit by bit: Social research in the digital age. Princeton University Press, Princeton, NJ
57. Lampe C, Ellison NB, Steinfield C (2008) Changes in use and perception of Facebook. In: Proceedings of the 2008 ACM conference on computer supported cooperative work, CSCW'08. New York, NY, pp 721–730
58. Liu Y, Kliman-Silver C, Mislove A (2014) The tweets they are a-changin': evolution of twitter users and behavior. In: Proceedings of the international AAAI conference on web and social media, Ann Arbor, MI

59. Danescu-Niculescu-Mizil C, West R, Jurafsky D, Leskovec J, Potts C (2013) No country for old members: user lifecycle and linguistic change in online communities. In: Proceedings of the 22nd international conference on world wide web,WWW'13. New York, NY, pp 307–318
60. Resnick P, Garrett RK, Kriplean T, Munson SA, Stroud NJ (2013) Bursting your (filter) bubble: strategies for promoting diverse exposure. In: Proceedings of the 2013 conference on computer supported cooperative work companion, CSCW'13. New York, NY, pp 95–100
61. Van Binh T, Minh D, Linh L, Van Nhan T (2023) Location-based service information disclosure on social networking sites: the effect of privacy calculus, subjective norms, trust, and cultural difference. Inf Serv & Use. 1–25
62. Newell ET, Dimitrov S, Piper A, Van Ruths D (2021) To buy or to read: how a platform shapes reviewing behavior. In: Proceedings of international conference on web and social media (ICWSM)
63. D'Alessio D, Allen M (2000) Media bias in presidential elections: a metaanalysis. J Commun 50:133–156
64. Blodgett SL, Green L, O'Connor B (2016) Demographic dialectal variation in social media: a case study of African-American English. In: Proceedings of the 2016 conference on empirical methods in natural language processing, Austin, TX, pp 1119–1130
65. Liang H, Fu K-W (2015) Testing propositions derived from twitter studies: generalization and replication in computational social science. PLoS ONE 10:e0134270
66. White RW (2016) Interactions with search systems. Cambridge University Press, Cambridge
67. Radford J, Joseph K (2020) Theory in, theory out: the uses of social theory in machine learning for social science. Front Big Data 3:18
68. Cerqueira V, Torgo L, Smailović J, Mozetič I (2017) A comparative study of performance estimation methods for time series forecasting. In: 2017 IEEE international conference on data science and advanced analytics (DSAA)8. IEEE, pp 529–53
69. Guest E, Vidgen B, Mittos A, Sastry N, Tyson G, Margetts H (2021) An expert annotated dataset for the detection of online misogyny. In: Proceedings of the 16th conference of the European chapter of the association for computational linguistics: main volume, Association for Computational Linguistics, pp 1336–1350
70. Agarwal P, Hawkins O, Amaxopoulou M, Dempsey N, Sastry N, Wood E (2021) Hate speech in political discourse: a case study of UK MPs on twitter. In: Proceedings of the 32nd ACM conference on hypertext and social media (HT '21). New York, NY, USA, pp 5–16
71. Zia HB, Raman A, Castro I, Anaobi IH, Cristofaro ED, Sastry N, Tyson G (2022) Toxicity in the decentralized web and the potential for model sharing. In: Proceedings of ACM measurement and analysis of computing system vol 6, 2, Article 35
72. Vidgen B, Thrush T, Waseem Z, Kiela D (2021) Learning from the worst: dynamically generated datasets to improve online hate detection. arXiv:2012.15761
73. Yin W, Agarwal V, Jiang A, Zubiaga A, Sastry N (2023) AnnoBERT: effectively representing multiple annotators' label choices to improve hate speech detection. Accepted In: The 17th international AAAI conference on web and social media (ICWSM)

The Theory of Fair Allocation Under Structured Set Constraints

Arpita Biswas, Justin Payan, Rik Sengupta, and Vignesh Viswanathan

Abstract The topic of fair allocation of indivisible items has been receiving significant attention because of its applicability in real-world settings, such as budgeted course allocations, room allocations, reviewer assignments, inheritance partitioning, and many others. In this chapter, we present an introduction to the theory of fairly allocating indivisible items in the presence of *structured* constraints, a natural setting that captures real-world restrictions. This setting has led to several works focusing on questions such as: (1) what are mathematically robust descriptions of these constraints? (2) how do we define appropriate fairness notions under these constraints? (3) do fair allocations always exist for all such relevant instances? (4) can we provide efficient algorithms to compute optimal or approximately optimal allocations? Typically, these questions become even more challenging when there are additional constraints on the feasible allocations, which may occur due to connectivity requirements, or incompatibility between certain item pairs. A large body of recent work has focused on defining appropriate fairness notions, providing existence guarantees, and handling computational issues in obtaining fair allocations under structured restrictions. We summarize this literature briefly.

Keywords Social choice theory · Fairness · Envy · Maximin · Indivisible items · Matroid · Connectivity · Graph theory

---

A. Biswas (✉)
School of Engineering and Applied Sciences, Harvard University, Cambridge, MA, USA
e-mail: arpitabiswas@seas.harvard.edu

J. Payan · R. Sengupta · V. Viswanathan
Department of Computer Science, University of Massachusetts Amherst, Amherst, MA, USA
e-mail: jpayan@cs.umass.edu

R. Sengupta
e-mail: rsengupta@cs.umass.edu

V. Viswanathan
e-mail: vviswanathan@cs.umass.edu

A. Mukherjee et al. (eds.), *Ethics in Artificial Intelligence: Bias, Fairness and Beyond*,
Studies in Computational Intelligence 1123,
https://doi.org/10.1007/978-981-99-7184-8_7

1 Introduction

Fairness is a fundamental consideration in many applications wherein a limited set of resources has to be allocated among multiple agents. The study of fair division in computer science, mathematics, and economics started with the formal introduction of the *cake-cutting problem*, where a single *divisible* item (such as a piece of land) needs to be *fairly* allocated among a set of agents. Since then, several fairness notions have been formalized and studied. The most well-studied notion among these is *envy-freeness* (EF), which requires ensuring that every agent values her allocated bundle as much as any other agent's allocated bundle, formally defined by Foley [25]. When there are two agents, the classical *cut-and-choose* protocol outputs an envy-free allocation—one agent (cutter) cuts the item into two pieces, and the other agent (chooser) chooses one of the pieces, leaving the cutter with the remaining piece. However, when there are three or more agents, finding an envy-free division is non-trivial. Stromquist [55] proposed the *moving-knife procedure*, which allocates connected envy-free pieces of the item to each agent. Since then, several works have proposed sophisticated methods to achieve envy-free allocations under various modeling assumptions; the books by Moulin [46] and Brandt et al. [16] provide excellent expositions.

In contrast, the study of fairness notions for the allocation of *indivisible* items is relatively recent, with Lipton et al. [43] and Budish [18] providing some of the early breakthroughs, primarily motivated by budgeted course allocations. The *indivisibility* assumption over the items is relevant for several real-world scenarios where the items cannot be fractionally allocated, or in other words, the items lose their values when broken into pieces. Unfortunately, fairness notions for the divisible setting may not always be applicable in the indivisible setting. For example, it is not possible to achieve an *envy-free* allocation when there is only one indivisible item and two agents—allocating the item to any one of the agents will induce envy in the other agent. This has led to a vast body of work where the primary interests are defining solution concepts that are appropriate and applicable for fairly allocating a set of indivisible items, establishing existential guarantees, and handling the computational issues surrounding the underlying solution concepts.

The focus on developing efficient solutions is motivated, in part, by websites such as *Spliddit* [28] (www.spliddit.org) and *Adjusted Winner* [15] (www.nyu.edu/projects/adjustedwinner/). Moreover, the solution concepts from the fair allocation of indivisible items have transcended their original boundaries and have been applied to define fairness in diverse application domains, such as envy-free tour-package recommendation [53], two-sided fair recommendations in e-commerce platforms [12, 49], envy-free classification [4], and preference-based loan sanctioning [60, 62].

The majority of work in the area of fair allocation has focused only on unconstrained settings. In practical applications, there may be restrictions on the group of items that are allowed to be allocated to any agent; for instance, cardinality constraints (there can be an upper bound on how many items from a particular category are assigned to an agent) or connectivity constraints (the items can be on a graph, with only connected sets of items allowed to be allocated to any agent). Moreover, there are situations where the agents are part of a social network (represented by a graph), necessitating the study of newer definitions of fairness and novel solutions. More recently, these constraints over items and agents have been investigated and formalized using mathematically robust descriptions of the constraints. Suksompong [57] provides a survey of fair division problems with constraints defined over items.

The literature on the problem of fair allocation of indivisible items is too extensive to exhaustively survey in this chapter. We summarize the existence and algorithmic results in indivisible settings where constraint structures are defined over items as well as over agents. We start with some basic definitions in Sect. 2. We then focus on three types of constrained settings, namely, matroid-constrained instances (Sect. 3), connectivity constraints over items (Sect. 4), and connectivity among agents (Sect. 5). Within each of these sections, we highlight a few open problems. We wrap up with a concluding discussion in Sect. 6.

2 Preliminaries and Notations

We are interested in the problem of fairly allocating a set of m *indivisible* items $\mathcal{M} = \{1, \ldots, m\}$ among a set of n agents $\mathcal{N} = \{1, \ldots, n\}$, whose preferences are known. The items lose their values when broken into pieces, and thus cannot be fractionally allocated to any agent. An *allocation*, $\mathbf{A}$, is an n-partition $\mathbf{A} = (A_1, A_2, \ldots, A_n) \in \Pi_n(\mathcal{M})$, where A_i is the subset of goods ("bundle") allocated to agent i and $\Pi_n(\mathcal{M})$ is the set of all n-partitions of $\mathcal{M}$. Note that because $\mathbf{A}$ is an n-partition, the following two properties hold:

(i) *Disjointness*: any single item is allocated to at most one agent, i.e., $A_i \cap A_j = \varnothing$ for any pair of agents $i, j \in \mathcal{N}$.
(ii) *Allocative efficiency*: no item remains unallocated, also known as *completeness*, i.e., $\cup_{i=1}^{n} A_i = \mathcal{M}$.

Valuation Functions
The *preference* or *valuation* of each agent i over each subset of items $S \subseteq \mathcal{M}$ is denoted by $v_i(S)$, where we can think of the valuations for agent i as a function $v_i : 2^{\mathcal{M}} \to \mathbb{R}$. In many scenarios, it is suitable to assume *additivity* of valuations, i.e., the value of a bundle is the sum of the values of all items in it: $v_i(S) = \sum_{g \in S} v_i(\{g\})$ for all $i \in \mathcal{N}$ and for all $S \subseteq \mathcal{M}$. Thus, instead of specifying preferences over subsets of items, it is enough to specify preferences over each individual item. For readability, we will refer to $v_i(\{g\})$ as $v_i(g)$.

Another common assumption on the valuation function is *monotonicity*: allocating additional items to an agent never makes her less happy than before. Under additive valuations, monotonicity is equivalent to having non-negative valuations for all the items. We call items as *goods* when $v_i(g) \geq 0$ for all $i \in \mathcal{N}$.

There are several results in the realm where all the items are considered to have non-positive valuations, i.e., $v_i(g) \leq 0$ for $g \in \mathcal{M}$, known as *chore* allocation. In this chapter, we use the terms *goods* and *chores* to indicate positively and negatively valued items respectively.

Fairness Notions

Two well-studied solution concepts for the problem of fairly allocating indivisible goods are *envy-freeness up to one good* (EF1; Definition 1) and *maximin fair share* (MMS; Definition 3). These notions were formally introduced by [18] as natural surrogates of the classical fairness notions for *divisible* goods, namely, *envy-freeness* (EF) [25, 55] (each agent values her bundle at least as much as anyone else's bundle) and *proportional fair share* (PFS) [54] (each agent receives a bundle worth at least $1/n$ times the total value of all goods in $\mathcal{M}$). The following definitions cover most of the fairness notions that we survey in this chapter:

Definition 1 *(Envy-Freeness up to k Goods (EFk))* An allocation $\mathbf{A} = (A_1, \ldots, A_n)$ is said to be *envy-free up to k goods* (EFk) if and only if, for every pair of agents $i, j \in \mathcal{N}$, there exists a subset of goods $\mathcal{K} \subseteq A_j$ with $|\mathcal{K}| \leq k$ such that $v_i(A_i) \geq v_i(A_j \setminus \mathcal{K})$. When $k = 1$, the allocation is *EF1*. When $k = 0$, the allocation is *envy-free* (EF).

Definition 2 *(Envy-Freeness up to the Least Valued Good (EFX)* [19]) An allocation $\mathbf{A} = (A_1, \ldots, A_n)$ is said to be *envy-free up to the least valued good* (EFX) if and only if, for every pair of agents $i, j \in \mathcal{N}$, the following holds:

$$v_i(A_i) \geq v_i(A_j \setminus \{g\}) \text{ for all goods } g \in A_j.$$

Definition 3 *(Maximin Share Allocation (MMS))* An allocation $(A_1, \ldots, A_n)$ is said to have *α-maximin share guarantee* (α-MMS) for some $\alpha \in (0, 1]$ if and only if, for all agents $i \in \mathcal{N}$, the following holds:

$$v_i(A_i) \geq \alpha \cdot \text{MMS}_i, \qquad \text{where} \quad \text{MMS}_i := \max_{\mathbf{B} \in \Pi_n(\mathcal{M})} \min_{j \in \mathcal{N}} v_i(B_j).$$

A 1-approximate maximin share allocation is an *MMS allocation*.

Scale of Fairness: There is a chain of implications between some of these fairness notions. For instance, EF $\Longrightarrow$ PFS $\Longrightarrow$ EFX $\Longrightarrow$ EF1 $\Longrightarrow$ EFk. Also, PFS $\Longrightarrow$ MMS $\Longrightarrow$ α-MMS (for $\alpha \in [0, 1]$). However, EF1 and MMS are incomparable [2].

In unconstrained settings, EF1 allocations always exist under monotone valuations. In fact, assuming additive valuations, a *round-robin* algorithm [19, 44] always returns an EF1 allocation. Moreover, an EF1 allocation can be computed even for general monotone valuations using the *cycle-elimination* algorithm by Lipton et al. [43]. However, the universal existence of EFX under additive valuations in the unconstrained setting remains open till now.

On the other hand, the existence of MMS allocations is not always guaranteed. [52] and [40] provide intricate counterexamples to refute the universal existence of

MMS allocations (even under additive valuations). [1, 52] propose algorithms to obtain a (2/3)-MMS allocation for additive valuations. Later, this guarantee was improved by Ghodsi et al. [27] to (3/4)-MMS.

Efficiency Criteria
The following definitions cover most of the allocative efficiency notions that we survey in this chapter:

Definition 4 (*Pareto Optimality (PO)*) An allocation $\mathbf{A} = (A_1, \ldots, A_n)$ *Pareto dominates* another allocation $\mathbf{A}' = (A'_1, \ldots, A'_n)$ if and only if $v_i(A_i) \geq v_i(A'_i)$ for all i, and $v_j(A_j) > v_j(A'_j)$ for some j. An allocation is *Pareto optimal* (PO) if no other allocation Pareto dominates it.

Definition 5 (*Nash Social Welfare (NSW)*) The *Nash social welfare* (or simply *Nash welfare*) (NSW) of an allocation $\mathbf{A} = (A_1, \ldots, A_n)$ is defined as $\left(\prod_{i \in \mathcal{N}} v_i(A_i)\right)^{1/n}$.

The allocation which maximizes the Nash welfare is often considered to be a reasonable trade-off between fairness and efficiency, and it is known to satisfy EF1 and PO under additive valuations [19]. However, as shown in Example 1, this need not be true in constrained settings.

3 Matroid-Constrained Fair Allocation

A *matroid* [48] is defined as a pair $(\mathcal{M}, \mathcal{I})$ where $\mathcal{M}$ is the ground set of elements, and $\mathcal{I}$ is a nonempty collection of subsets of $\mathcal{M}$ (referred to as the set of *independent sets*), satisfying:

1. Hereditary property: if $B \in \mathcal{I}$ and $A \subset B$, then $A \in \mathcal{I}$, and
2. Independent Set Exchange: if $A, B \in \mathcal{I}$ and $|A| < |B|$, then there exists an element $x \in B \setminus A$ such that $A \cup \{x\} \in \mathcal{I}$.

Matroids provide a framework for representing combinatorial constraints that allow the formulation of a broader set of interesting and challenging practical problems in the space of fair allocation.

An instance of matroid-constrained fair allocation is given as a 4-tuple $\langle \mathcal{M}, \mathcal{N}, (v_i)_{i \in \mathcal{N}}, \mathcal{F} \rangle$ where $\mathcal{M}, \mathcal{N}, v_i$ are defined as before, and $\mathcal{F}$ denotes the set of all *matroid-feasible* allocations, whose constituent bundles are independent sets of a given matroid $(\mathcal{M}, \mathcal{I})$, i.e., $\mathcal{F} := \{\mathbf{A} = (A_1, \ldots, A_n) \in \Pi_n(\mathcal{M}) : A_i \in \mathcal{I}, \forall i \in \mathcal{N}\}$.

Matroids have been considered by Gourvès and Monnot [29] and Gourvès et al. [30] as an admissibility criterion to the fair allocation problem. In particular, their problem formulation requires that the *union* of all the allocated goods is an independent set as well. They provide a (1/2)-MMS algorithm for any number of agents, a $(1 - \varepsilon)$-MMS algorithm for two agents, and an $(8/9 - \varepsilon)$-MMS algorithm for three agents. However, their algorithms do not necessarily result in *complete* allocations of all goods. In contrast, there are a few works that aim at ensuring allocative efficiency (completeness) while fairly allocating goods under matroid constraints, which we survey next.

Partition Matroid

In the partition matroid-constrained setting [10], the indivisible goods are categorized into disjoint groups and an upper limit is specified on the number of goods that can be allocated from each category to any agent. Biswas and Barman [10] show that (1/3)-MMS can be obtained efficiently using the fact that the problem can be reduced to an unconstrained fair allocation problem with submodular valuations (matroid rank functions) and then use a *bag-filling algorithm* [27]—where a partial bundle B is being filled unless for some agent i, $v_i(B) \geq 1/3\text{-MMS}_i$. This was later improved by Hummel and Hetland [36] to (1/2)-MMS by using a modification of the same algorithm. The initialization takes care of the partition matroid feasibility where $|C_h/n|$ least-valuable available goods in each category C_h are added to the bundle. This algorithm also obtains a 2-MMS allocation for chore division in the constrained setting.

Biswas and Barman [10] provide an EF1 solution that iteratively allocates goods from each category in a round-robin manner. After allocating all goods from one category, the resulting envy-graph of the agents (introduced by Lipton et al. [43]) is updated by cycle elimination, followed by topologically sorting the acyclic envy-graph to determine the round-robin ordering for the next category. Interestingly, this algorithm also guarantees 1-MMS when the valuations are binary (i.e., $v_i(g) \in \{0, 1\}$).

The non-existence of EFX can be shown using a counterexample [10] with 2 agents and 4 goods, for a uniform matroid, i.e., a single category C_1 of goods with upper limit $k_1 = 2$. Let the valuations of both the agents be identical and as follows: $v(g_1) = 50$ and $v(g_2) = v(g_3) = v(g_4) = 1$. Then, this instance admits no feasible EFX allocation. In fact, this counterexample refutes the existence of another fairness notion, called *envy-freeness up to one less preferred good* (EFL) [5], in the cardinality-constraint setting.

Laminar and Base-Ordered Matroids

For more general (e.g., laminar and base-ordered) matroids, Biswas and Barman [11] provide an algorithm, called SWAP, that computes an EF1 allocation under identical valuations (i.e., for any $i, j \in \mathcal{N}$, $v_i(g) = v_j(g) = v(g)$ for each $g \in \mathcal{M}$). The algorithm initializes an allocation by computing a matroid-feasible partition (using a method proposed by Gabow and Westermann [26]) and then iteratively reallocates goods between the bundles until an EF1 allocation is obtained. The reallocation strategy maintains matroid feasibility at each iteration and ensures polynomial-time convergence. For non-identical valuations, Dror et al. [22] show that EF1 allocations can be obtained for base-ordered matroid-constrained instances with at most 3 agents having non-identical binary valuations using the SWAP algorithm. The resulting allocation is also a social welfare maximizer and thus Pareto optimal (PO) under the constrained setting. The algorithm by Li and Vetta [42] that establishes the existence of (11/30)-MMS under hereditary set constraints can be adapted to show the

Table 1 Summary of results on the matroid-constrained fair allocation of indivisible goods. All valuations are assumed to be additive

Matroid	Agents	Fairness	Efficiency	Exists	References
Partition	n	EF1	–	Yes	[6]
	2	EF1	–	Yes	[22]
	n, binary	EF1	PO	Yes	[22]
	2, identical	EFX	–	No	[6]
	n	1-MMS	–	No	[40]
	n	$\frac{1}{2}$-MMS		Yes	[36]
Laminar and Base-Ordered	n, identical	EF1	–	Yes	[11]
	≤ 3, binary	EF1	PO	Yes	[22]
	n	$\frac{11}{30}$-MMS	–	Yes	[42]

same approximation factor for achieving MMS guarantees under non-identically valued matroid constraints.

Note that, under matroid constraints, the NSW-maximizing allocation need not be EF1 when agents' valuations are non-identical (see Example 1).

Example 1 Let $\mathcal{N} = \{a_1, a_2\}$ and $\mathcal{M} = \{g_1, g_2, g_3, g_4\}$ with valuations as follows:

	g_1	g_2	g_3	g_4
a_1	1	1	$2+\varepsilon$	$2+\varepsilon$
a_2	0	0	5	5

The allocation $A_1 = \{g_1, g_2\}$ and $A_2 = \{g_3, g_4\}$ maximizes NSW giving a value $2\sqrt{5}$ under the uniform matroid constraint with $k = 2$. However, the allocation (A_1, A_2) is not EF1. Any EF1 allocation assigns an agent exactly one good from $\{g_1, g_2\}$ and exactly one good from $\{g_3, g_4\}$. The Nash Welfare value of such an allocation is $\sqrt{(3+\varepsilon) \times 5} < 2\sqrt{5}$ for $\varepsilon \in (0, 1)$. This shows that the NSW maximizer may not be EF1 under matroid constraints (Table 1).

Open Question 1 Do EF1 allocations always exist for general matroid-constrained fair allocation instances (e.g., graphic and linear matroids), under identical additive valuations, even with 2 agents?

Open Question 2 Do EF1 allocations always exist for any matroid-constrained fair allocation instance with $n > 3$ agents under non-identical additive valuations?

Open Question 3 Can we efficiently obtain an α-MMS allocation where $\alpha \in (0.5, 1)$ for $n > 2$ agents, under non-identical additive valuations?

Other Cardinality Constraints
Ferraioli et al. [24] consider fair allocation problems where each agent must receive exactly k goods, for a given integer k. Their fairness objective is to maximize the utility of the least happy agent (also known as *maximin* or *Santa Claus problem*), and they provide a $(1/k)$-approximate solution. Payan and Zick [50] study the problem of obtaining EF1 allocation among conference papers. They use a modified round-robin procedure that satisfies cardinality constraints (each paper needs exactly k reviewers) with a partition matroid constraint (a paper can be assigned to a given reviewer at most once). Moreover, there are several recent approaches where matroid structures are assumed to be captured within the valuation function by means of a *matroid rank function* that leads to a special case of submodular valuations; however, the solutions do not guarantee complete allocations and therefore are beyond the scope of this survey.

4 Connectivity Constraints on Goods

One natural and widely used set of constraints is modeled by a graph on the set of items $\mathcal{M}$. In this setting, we are given an undirected graph $G = (\mathcal{M}, E)$ in addition to $\mathcal{N}$, $\mathcal{M}$, and v_i defined as before. A valid allocation $\mathbf{A} = (A_1, A_2, \ldots A_n)$ is again an n-partition of $\mathcal{M}$ that must have the property that for any $i \in \mathcal{N}$, A_i induces a connected subgraph in G. By defining the appropriate graphs, this connectivity constraint captures many natural scenarios—allocation of connected plots of land, consecutive time slots in a shared computing environment, and desks in a shared office where teams prefer to sit together, among many others. Most of the works cited in this section assume additive valuations, though there are some results beyond additivity.

Achieving Fairness under Connectivity Constraints
Bouveret et al. [13] was the first to introduce graph connectivity constraints in the allocation of indivisible items. They adapt the classical *moving-knife* procedure from divisible good allocation in the indivisible setting, and demonstrate that when G is a tree, an MMS allocation always exists and can be found in polynomial time. The existence of an MMS allocation extends to agents with arbitrary monotonic utility functions on any acyclic graph [57]. Furthermore, there is an instance on an 8-cycle that does not admit an MMS allocation. In contrast, the problem of deciding whether stronger fairness, such as EF or PFS, exists is NP-hard even for paths. However, approximate versions of EF and PFS can be obtained in polynomial time [56, 61]. When the underlying graph is a star, deciding whether an EF allocation exists remains NP-hard, but PFS is decidable in polynomial time. The same paper goes on to show algorithmic approaches for certain particularly simple graphs, obtaining FPT (fixed-parameter tractable) and XP (slice-wise polynomial) algorithms parameterized by the number of types of agents to obtain PFS and EF allocations, respectively. [14] extended the analysis to chores (Table 2).

Table 2 Summary of key results for connectivity constraints on goods. The **Exists** column describes if the fairness notion can always be satisfied. If yes, the **Complexity** listed is the complexity of computing a satisfying allocation. Otherwise, it is the complexity of deciding if the notion can be satisfied for a given instance

Graph	Valuations	Fairness	Exists	Complexity	References
Trees	Additive	MMS	Yes	P	[13]
$\mathsf{treewidth}(G) = 2$	Additive	MMS	No	NP-hard	[32]
Path	Additive	EF	No	NP-hard	[13]
	Additive	PFS	No	NP-hard	[13]
	Monotone	EF1	Yes	Open	[38]
Star	Additive	EF	No	NP-hard	[13]
	Binary	EF1/EFX	No	NP-hard	[20]
	Additive	PFS	No	P	[13]

The natural setting where the underlying graph is a path captures scheduling constraints, in which agents prefer to receive contiguous time blocks. Biló et al. [9] show EF1 allocations exist when agents' valuations are identical, and EF1 allocations on paths exist for up to 4 agents with monotonic valuations. They also characterize the set of graphs for which EF1 allocations exist when there are only 2 agents. There is an EF1 allocation if and only if the constraint graph G admits a *bipolar ordering*—an ordering of the vertices in which every prefix and every suffix of the ordering is connected. Igarashi [38] extends the EF1 existence on a path to n agents with monotonic valuations. The main technical proofs in both of these papers rely on a rounding technique for a particular simplex defined over tuples of bundles. Igarashi [38] was able to extend the techniques from 4 to n agents by breaking the symmetry of this simplex cleverly. In fact, this result generalizes to any graph with a Hamiltonian path. Because the construction is exponential time, the complexity of EF1 allocation on paths and graphs with Hamiltonian paths remains unknown.

Open Question 4 What is the time complexity of computing an EF1 allocation over path connectivity constraints, even under binary additive valuations?

Open Question 5 Are there more general graphical connectivity constraints that always admit EF1 under some set of valuations? Does there exist a counterexample for every graph with no Hamiltonian path under monotone valuations?

Truszczynski and Lonc [59] investigate the complexity of determining if an MMS allocation exists when goods are arranged in a cycle. They demonstrate that the value of the MMS share for each agent is computable in polynomial time in this case, which implies that the decision problem for MMS is in NP. Greco and Scarcello [32] show that the computational problem of finding optimal MMS allocations is intractable, even in instances where they are guaranteed to exist. This intractability holds even when the agents have identical valuation functions, or when the underlying graph has treewidth 2 (which includes cycles). If we only wish to satisfy approximate

MMS, $\frac{\sqrt{5}-1}{2}$-MMS is achievable for cycle constraint graphs under additive valuations [59]. Moreover, if the agents have identical valuations *and* the graph has bounded treewidth, there is a polynomial time algorithm for computing optimal MMS allocations, using an optimal tree decomposition and a nondeterministic logspace protocol that can be converted to a polynomial time algorithm [32].

On general graphs, Deligkas et al. [20] give XP algorithms for PFS, EF, EFX, and EF1, parameterized by the clique-width and the number of agent types, as well as by the treewidth and number of agent types. They additionally show that EF1 and EFX are NP-hard on stars. Deligkas et al. [21] characterize the computational complexity of different variants of the graph cutting problem, showing NP-completeness for finding an EF allocation for two agents even when the graph is as simple as two vertices plus a matching. When the number of edges in the graph is constant, however, they show a polynomial time algorithm for finding an EF allocation, that involves linear programming subroutines and multidimensional geometry.

Trade-Offs under Connectivity Constraints

Imposing connectivity constraints can create trade-offs with either the fairness requirement or the efficiency, which we can measure by means of notions such as the *price of fairness*[1], *price of connectivity*[2], and other similar metrics of measuring inherent trade-offs. Bei et al. [7] show, for two agents, the connectivity constraint allows for (3/4)-MMS allocations when the graph G is 2-connected, and (1/2)-MMS otherwise. When the vertex connectivity is 1, the price of connectivity for MMS is equal to the maximum number of connected components formed by deleting a vertex. Furthermore, the price of connectivity for MMS is upper-bounded by 4/3 for all graphs. They extend the results of Biló et al. [9], determining the smallest k for which an EFk allocation exists on various graphs. The value of k is tied to the connectivity of the graph; for less connected graphs, the smallest possible value of k is higher.

Igarashi and Peters [39] dive deeper into the trade-off between allocative efficiency and connectivity, showing that PO allocations are computable in poly-time on paths and stars, but are NP-hard on trees. The universal existence of MMS with tree-structured constraints extends to PO and MMS allocations, but such an allocation is NP-hard to compute. They also show the incompatibility of PO and EF1 under path constraints, which contrasts with the celebrated result of Caragiannis et al. [19] that the maximum Nash welfare solution is PO and EF1 in the unconstrained regime. Sun and Li [58] restrict their analysis to the price of fairness for α-MMS and PFS1 over goods connected on a path in terms of various notions of welfare.

Höhne and van Stee [37] consider the setting of chore allocation on a path graph. They show the price of fairness is roughly comparable for goods and chores for PFS, but not for most other standard fairness criteria.

Open Question 6 What is the price of fairness for chore division on more general graphs than paths?

[1] The *price of fairness* is the ratio of the welfare of an optimal allocation to the welfare of an optimal fair allocation for goods, and the inverse of this ratio for chores.

[2] The *price of connectivity* is the ratio of the optimal criterion value with the connectivity constraint enforced, to the optimal criterion value without connectivity enforced.

5 Connectivity on Agents

Some recent works considered fair allocations among *agents* who are placed at the vertices of a graph. In this setting, we are given $\mathcal{N}$, $\mathcal{M}$, and v_i as usual, together with an undirected graph $G = (\mathcal{N}, E)$ representing a social network on the agents. In this scenario, the fairness constraints are required to be maintained only along the edges of the agent graph G.

Single-Item Bundles

One particular case of fair allocation on graphs that has been considered recently is *house allocation* – fair allocation where agents must receive exactly one item each, and typically, $|\mathcal{N}| = |\mathcal{M}|$.

Beynier et al. [8] studied *local envy-freeness* in house allocation, the problem of checking the existence of an allocation while ensuring that no envy is present along any edge of the underlying graph. They characterize the computational complexity of this problem with respect to various natural graph parameters such as maximum and minimum degree, number of disjoint cliques, and size of minimum vertex cover. Notably, their model involves agents with *possibly distinct* ordinal preferences, and they only consider envy-free allocations, instead of trying to minimize envy in instances when envy-freeness is not possible.

Hosseini et al. [33] consider the problem of minimizing envy on arbitrary graphs when agents have identical valuations. They show that the problem is NP-complete even when the underlying graph is a disjoint union of paths, cycles, or stars, even though they each admit FPT algorithms in the number of connected components in the graph.

Other works seek to obtain envy-free allocations, Pareto optimality, or other objectives in the house allocation setting by swapping single items along a graphical structure [31, 35, 41, 47]. A typical model used in these works is to consider a *social trading graph*, where pairs of agents interact sequentially and offer to swap their current bundles with each other as long as some mutual incentive is present.

Open Question 7 What is the complexity of computing an allocation that minimizes the maximum envy that agents feel toward their neighbors?

Parameterized Complexity of Envy-Freeness

Another line of work studies the complexity of computing locally envy-free allocations (summarized in Table 3). Bredereck et al. [17] present fixed-parameter tractability results, mainly parameterized by the number of agents, though they leave results using graph structure for future work. Their FPT algorithms rely on constructing complex gadget graphs from the instance graphs and extending partial allocations item by item. Eiben et al. [23] extend these results, showing a number of parameterized complexity results relating the treewidth, clique-width, number of agent types, and number of item types to the complexity of determining if an envy-free allocation exists on a graph. Misra and Nayak [45] study the complexity of locally fair and Pareto optimal allocations when agents have binary valuations. Specifically,

Table 3 Summary of results on Local Envy-Freeness (LEF)

Valuations	Fairness	Parameter	Complexity	References
Binary	LEF	–	NP-hard	Bredereck et al. [17]
	LEF, PO	Vertex cover	W[1]-hard	Misra and Nayak [45]
Additive	LEF	Treewidth	XP	Eiben et al. [23]
		Cliquewidth	XP	Eiben et al. [23]
		Treewidth	W[1]-hard	Eiben et al. [23]
		Agents	W[1]-hard	Bredereck et al. [17]
		Goods	W[1]-hard	Bredereck et al. [17]

Table 4 Summary of results on Envy-Freeness Relaxations over Graphs

Graph	Valuation	Fairness	Exists	References
Arbitrary graph	General	EEF	–	[3]
(Generalized) Star	General	EFX	yes	[51]
(Generalized) Path	General	EFX	yes	[51]
K_4	Lexicographic	EFX	no	[34]
$\mathsf{diameter}(G) \geq 4$	Lexicographic	EFX	yes	[51]

they study the complexity of computing locally envy-free and local PFS allocations parameterized by the number of agents, number of goods, and the vertex cover of the social network.

Envy-Freeness Relaxations over Graphs

The computational intractability of finding envy-free allocations has led to several works studying the relaxations of envy-freeness over graphs (summarized in Table 4). Payan et al. [51] show that stars, short paths, and their generalizations admit allocations that satisfy the EFX criterion on every edge of the underlying graph. The problem becomes harder with both goods and chores. However, even though Hosseini et al. [34] show that EFX allocations of goods and chores do not exist in general even for the restricted class of *lexicographic* valuation functions, Payan et al. [51] demonstrate an EFX allocation of goods and chores on any graph with diameter 4 or more.

Open Question 8 Which graph structures admit EFX allocations under general (or additive) valuations?

6 Conclusion

In this chapter, we survey the problem of fair allocation of indivisible items under three structured set constraints, such as matroids defined over items, connectivity constraints over goods, and network structures over agents. We highlight important results covering the existence and hardness results for a few common fairness notions, elaborate on common algorithmic tools, and list several open questions in the existing line of work. In summary, even though it is possible to guarantee a few types of fairness under constrained settings, a procedure to find such allocations is non-trivial in most cases and sometimes computationally intractable. Given the wide spectrum of real-world problems that can be modeled using structured constraints, (e.g., allocation of scarce healthcare resources or course allocation), we hope that this survey inspires exciting new questions and further research.

References

1. Amanatidis G, Markakis G, Nikzad A, Saberi A (2015) Approximation Algorithms for computing maximin share allocations. In: Proceedings of the 42nd ICALP, pp 39–51
2. Aziz H, Rauchecker H, Schryen G, Walsh T (2017) Approximation algorithms for max-min share allocations of indivisible chores and goods. In: Proceedings of the 31st AAAI, pp 335–341
3. Aziz H, Bouveret S, Caragiannis I, Giagkousi I, Lang J (2018) Knowledge, fairness, and social constraints. In: Proceedings of the 32nd AAAI, pp 4638–4645
4. Balcan M-FF, Dick T, Noothigattu R, AD (2019) Procaccia: envy-free classification. In: Proceedings of the 33rd NeurIPS, pp 1238–1248
5. Barman S, Biswas A, Murthy SKK, Narahari Y (2018a) Groupwise maximin fair allocation of indivisible goods. In: Proceedings of the 32nd AAAI, pp 917–924
6. Barman S, Krishnamurthy SK, Vaish R (2018b) Finding fair and efficient allocations. In: Proceedings of the 18th EC, pp 557–574
7. Bei X, Igarashi X, Lu X, Suksompong W (2021) The price of connectivity in fair division. In: Proceedings of the 35th AAAI, pp 5151–5158
8. Beynier A, Chevaleyre Y, Gourvès L, Lesca J, Maudet N, Wilczynski A (2018) Local envy-freeness in house allocation problems. In: Proceedings of the 17th AAMAS, pp 292–300
9. Bilò V, Caragiannis I, Flammini M, Igarashi A, Monaco G, Peters D, Vinci C, Zwicker WS (2022) Almost envy-free allocations with connected bundles. Games Econom Behav 131:197–221
10. Biswas A, Barman S (2018) Fair division under cardinality constraints. In: Proceedings of the 27th IJCAI, pp 91–97
11. Biswas A, Barman S (2019) Matroid constrained fair allocation problem. In: Proceedings of the 33rd AAAI, pp 9921–9922
12. Biswas A, Patro GK, Ganguly N, Gummadi KP, Chakraborty A (2022) Toward fair recommendation in two-sided platforms. ACM Trans Web (TWEB) 16(2):1–34
13. Bouveret S, Cechlárová K, Elkind E, Igarashi A, Peters D (2017) Fair division of a graph. In: Proceedings of the 26th IJCAI, pp 135–141
14. Bouveret S, Cechlarova K, Lesca J (2019) Chore division on a graph. Auton Agent Multi-Agent Syst 33(5):540–563
15. Brams SJ, King DL (2005) Efficient fair division: help the worst off or avoid envy? Ration Soc 17(4):387–421

16. Brandt F, Conitzer V, Endriss U, Lang J, Procaccia AD (eds) (2016) Handbook of computational social choice. Cambridge University Press
17. Bredereck R, Kaczmarczyk A, Niedermeier R (2022) Envy-free allocations respecting social networks. Artif Intell 305
18. Budish E (2011) The combinatorial assignment problem: approximate competitive equilibrium from equal incomes. J Polit Econ 119(6):1061–1103
19. Caragiannis I, Kurokawa D, Moulin H, Procaccia AD, Shah N, Wang J (2019) The unreasonable fairness of maximum Nash welfare. ACM Trans Econ Comput (TEAC) 7(3):1–32
20. Deligkas A, Eiben E, Ganian R, Hamm T, Ordyniak S (2021a) The parameterized complexity of connected fair division. In: Proceedings of the 30th IJCAI, pp 139–145
21. Deligkas A, Eiben E, Ganian R, Hamm T, Ordyniak S (2021b) The complexity of envy-free graph cutting
22. Dror A, Feldman M, Segal-Halevi E (2021) On fair division under heterogeneous matroid constraints. In: Proceedings of the 35th AAAI, pp 5312–5320
23. Eiben E, Ganian R, Hamm T, Ordyniak S (2020) Parameterized complexity of envy-free resource allocation in social networks. In: Proceedings of the 34th AAAI, pp 7135–7142
24. Ferraioli D, Gourvès L, Monnot J (2014) On regular and approximately fair allocations of indivisible goods. In: Proceedings of the 13th AAMAS, pp 997–1004
25. Foley D (1967) Resource allocation and the public sector. Yale Econ Essays 7(1):73–76
26. Gabow HN, Westermann HH (1992) Forests, frames, and games: algorithms for matroid sums and applications. Algorithmica 7(1–6):465
27. Ghodsi M, Hajiaghayi MT, Seddighin M, Seddighin S, Yami K (2018) Fair allocation of indivisible goods: improvements and generalizations. In: Proceedings of the 19th EC, pp 539–556
28. Goldman J, Procaccia AD (2015) Spliddit: unleashing fair division algorithms. ACM SIGecom Exchanges 13(2):41–46
29. Gourvès L, Monnot J (2017) Approximate maximin share allocations in matroids. In: Proceedings of the 10th CIAC, pp 310–321
30. Gourvès L, Monnot J, Tlilane L (2014) Near fairness in matroids. In: Proceedings of the 21st ECAI, pp 393–398
31. Gourvès L, Lesca J, Wilczynski A (2017) Object allocation via swaps along a social network. In: Proceedings of the 26th IJCAI, pp 213–219
32. Greco G, Scarcello F (2020) The complexity of computing maximin share allocations on graphs. In: Proceedings of the 34th AAAI, pp 2006–2013
33. Hosseini H, Payan J, Sengupta R, Vaish R, Viswanathan V (2023a) Graphical house allocation. In: Proceedings of the 22nd AAMAS
34. Hosseini H, Sikdar S, Vaish R, Xia L (2023b) Fairly dividing mixtures of goods and chores under lexicographic preferences. In: Proceedings of the 22nd AAMAS
35. Huang S, Xiao M (2019) Object reachability via swaps along a line. In: Proceedings of the 33rd AAAI, pp 2037–2044
36. Hummel H, Hetland ML (2022) Guaranteeing half-maximin shares under cardinality constraints. In: Proceedings of the 21st AAMAS, pp 1633–1635
37. Höhne F, van Stee R (2021) Allocating contiguous blocks of indivisible chores fairly. Inf Comput 281
38. Igarashi A (2022) How to cut a discrete cake fairly
39. Igarashi A, Peters D (2019) Pareto-optimal allocation of indivisible goods with connectivity constraints. In: Proceedings of the 33rd AAAI, pp 2045–2052
40. Kurokawa D, Procaccia AD, Wang J (2016) When can the maximin share guarantee be guaranteed? In: Proceedings of the 30th AAAI, pp 523–529
41. Li F, Zheng X (2021) Maximum votes pareto-efficient allocations via swaps on a social network. In: Proceedings of the 46th MFCS
42. Li Z, Vetta A (2018) The fair division of hereditary set systems. In: Proceedings of the 14th WINE, pp 297–311

43. Lipton RJ, Markakis E, Mossel E, Saberi A (2004) On approximately fair allocations of indivisible goods. In: Proceedings of the 5th EC, pp 125–131
44. Markakis E (2017) Approximation algorithms and hardness results for fair division with indivisible goods. In: Trends in computational social choice, pp 231–247
45. Misra N, Nayak D (2022) On fair division with binary valuations respecting social networks. In: Proceedings of the 8th CALDAM, pp 265–278
46. Moulin H (2003) Fair division and collective welfare. MIT Press
47. Müller L, Bentert M (2021) On reachable assignments in cycles. In: Proceedings of the 7th ADT, pp 273–288
48. Oxley JG (1992) Matroid theory. Oxford University Press
49. Patro GK, Biswas A, Ganguly N, Gummadi KP, Chakraborty A (2020) FairRec: two-sided fairness for personalized recommendations in two-sided platforms. In: Proceedings of the WWW, pp 1194–1204
50. Payan J, Zick Y (2022) I will have order! optimizing orders for fair reviewer assignment . In: Proceedings of the 31st IJCAI, pp 440–446
51. Payan J, Sengupta R, Viswanathan V (2023) Locally EFX allocations over a graph. In: Proceedings of the 22nd AAMAS (Extended Abstract)
52. Procaccia AD, Wang J (2014) Fair enough: guaranteeing approximate maximin shares. In: Proceedings of the 15th EC, pp 675–692
53. Serbos D, Qi S, Mamoulis N, Pitoura E, Tsaparas P (2017) Fairness in package-to-group recommendations. In: Proceedings of the WWW, pp 371–379
54. Steinhaus H (1948) The problem of fair division. Econometrica 16:101–104
55. Stromquist W (1980) How to cut a cake fairly. Am Math Mon 87(8):640–644
56. Suksompong W (2019) Fairly allocating contiguous blocks of indivisible items. Discret Appl Math 260:227–236
57. Suksompong W (2021) Constraints in fair division. ACM SIGecom Exchanges 19(2):46–61
58. Sun A, Li B (2022) On the price of fairness of allocating contiguous blocks
59. Truszczynski M, Lonc Z (2020) Maximin share allocations on cycles. J Artif Intell Res 69:613–655
60. Ustun B, Liu Y, Parkes D (2019) Fairness without harm: decoupled classifiers with preference guarantees. In: Proceedings of the 36th ICML, pp 6373–6382
61. Yang M (2022) Fairly allocating (contiguous) dynamic indivisible items with few adjustments
62. Zafar MB, Valera I, Gomez Rodriguez M, Gummadi KP (2017) Fairness beyond disparate treatment & disparate impact: learning classification without disparate mistreatment. In: Proceedings of the WWW, pp 1171–1180

Interpretability of Deep Neural Models

Sandipan Sikdar and Parantapa Bhattacharya

Abstract The rise of deep neural networks in machine learning has been remarkable, leading to their deployment in algorithmic decision-making. However, this has raised questions about the explainability and interpretability of these models, given their growing importance in society. To address this, the field of interpretability in machine learning has been developed, with the goal of creating frameworks that can explain the decisions of a machine learning model in a way that is comprehensible to humans. This could be essential in building trust in the system, as well as debugging models for potential errors and meeting legal requirements (e.g., GDPR). Even though the success of deep neural network is attributed to its ability to capture higher level feature interactions, most of existing frameworks still focus on highlighting important individual features (e.g., words in text or pixels in images). Hence, to further improve interpretability, we propose to quantify the importance of feature interactions in addition to individual features. In this work, we introduce integrated directional gradients (IDG), a game-theory inspired method for assigning importance scores to higher level feature interactions. Our experiments with DNN-based text classifiers on the task of sentiment classification demonstrate that IDG is able to effectively capture the importance of feature interactions.

1 Introduction

Deep Neural Networks (DNN) [19] have been immensely successful in a variety of tasks across different domains like computer vision, natural language processing in a variety of tasks like image recognition, and natural language understanding among others. Much of this success can be attributed to their increasing complexity which

S. Sikdar (✉)
L3S Research Center, Leibniz University Hannover, Hanover, Germany
e-mail: sandipan.sikdar@l3s.de

P. Bhattacharya
Biocomplexity Institute, University of Virginia, Charlottesville, USA
e-mail: sandipan.sikdar@l3s.de

A. Mukherjee et al. (eds.), *Ethics in Artificial Intelligence: Bias, Fairness and Beyond*,
Studies in Computational Intelligence 1123,
https://doi.org/10.1007/978-981-99-7184-8_8

has allowed them to "learn" more and more nuances of the data that they take as input.

Recently, the community has started developing strategies for explaining the output of DNNs for a given input. These strategies involve the use of feature attribution scores or saliency maps [35, 42] to provide an understanding of the black box model. Numerous studies have been conducted to determine the most influential features of a given input. However, the modern DNNs tend to learn from higher order feature interactions rather than just individual features. Hence, the focus of explainability methods have shifted toward considering feature interactions instead of individual features [3, 4, 15, 40, 43].

On the other hand, as has been pointed out in the existing literature [42], it is objectively difficult to compare two attribution methods based on the provided attribution scores. More importantly, when an attribution method generates non-intuitive results, it is unclear if the output was a result of (i) discrepancies in the explanation method, (ii) discrepancies in the DNN model itself or (iii) discrepancies of the training data. Hence, following previous literature [3, 42, 43], we propose an axiomatic approach. More specifically, we formally define the problem of feature interaction attribution (interchangeably referred to as feature group attribution) and then propose a set of axioms/properties which any solution to this problem should satisfy.

As a solution, we introduce integrated directional gradients (IDG) which are inspired by cooperative game theory. IDG explores axioms satisfied by a *well-behaved* characteristic function in addition to the solution concepts. In specific, we design our characteristic function in a way that it satisfies a set of axioms (Sect. 3.2) to ensure well-behavedness. We observe that well-behaved characteristic functions allow for defining simpler and more intuitive axioms.

We deploy IDG to state-of-the-art text classifiers based on contextual word embedding models such as BERT [6] and XLNet [44]. We illustrate the nature of explanations and attributions computed with IDG using an example in Fig. 1a. We also demonstrate the discrepancy in attribution score when the scores of the individual words are added to obtain the attribution score of a phrase (Fig. 1b). As input, IDG additionally feature groups structured in a hierarchical way (Sect. 3).

Note that the IDG framework along with a part of the results discussed here has already been presented in [36]. However, in this article, we present some interesting extensions. Firstly, we provide a detailed overview of the state-of-the-art interpretability methods, particularly in the domain of natural language processing. Secondly, we also provide additional results on evaluating interpretability methods through "CHECKLIST" tests and natural language generation. The rest of the article is organized as follows—in the following section, we provide a detailed overview of the interpretability methods developed in the context of natural language processing. We then formally define the feature group attribution problem and briefly describe the proposed method IDG. In Sect. 4, we describe the evaluation setup and present the key results. We conclude by summarizing our work in Sect. 5.

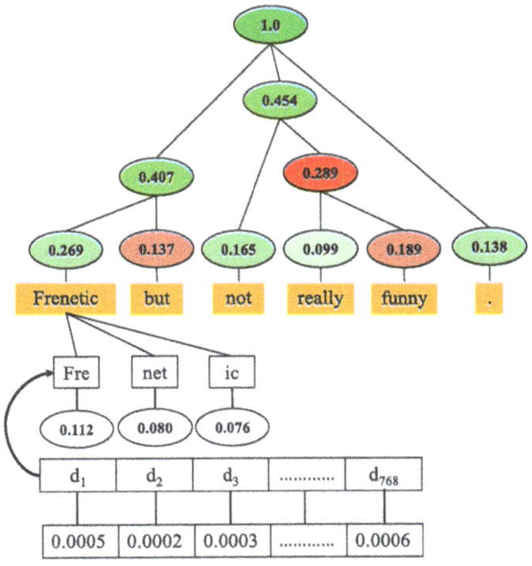

(a)

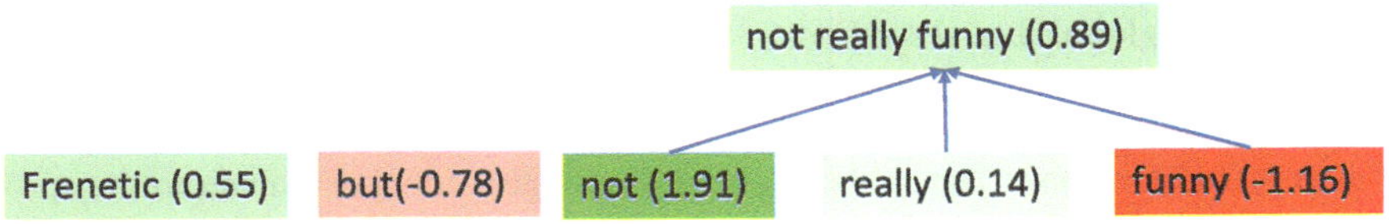

(b)

Fig. 1 **a** Computing attribution score (value function v). The task is binary sentiment classification. Red and green respectively denote negative or positive contributions to the model output while the color intensity represents the level of importance. **b** Discrepancy in the attribution score when simply adding the attribution scores of the words to obtain attribution scores of phrases. We use integrated gradients (IG) to compute the attribution scores of individuals and then sum them to the attribution score. Ideally, we would expect the attribution score of the phrase `not really funny` to be greater than the scores of the individual words. However, it is not so

2 Background and Related Work

Interpretability and explanability: In this article, we use the terms interpretability and explanability interchangeably. However, recent studies such as [32] propose to differentiate between the terms explainabilty and interpretability. In this context, a model is interpretable if it generates human understandable explanations on its own. In contrast, an explainable model requires a separate framework to generate explanations. However, such terminology is still not standardized. In our setting, we assume the model at hand to be a black box and our goal is to design a framework that generates explanations for the outputs obtained.

Several diverse methods have been proposed to explain inference results of deep neural network architectures [1, 16, 24, 30, 33, 42]. These methods could be categorized in several ways. However, we will, in part, follow the categorizations proposed in [2].

Post-hoc versus self-explanatory: Post-hoc methods such as [30, 35, 42] aim to explain already trained machine learning models. For example, given a target model and input, integrated gradients [42], assigns importance scores to each individual feature of the input. Similarly, LIME [30] learns an interpretable surrogate model (e.g., linear regression) around the neighborhood of the prediction. Once trained, the surrogate model is used to explain the decision. In self-explanatory methods [17, 28, 29], an additional explanation module is already included in the network architecture that generates explanations. The explanation module is trained jointly with the predictor module and may [11, 28] or may not [20, 45] require supervision.

Model-specific versus model-agnostic: As the name suggests, model-specific methods can only be deployed to particular neural network architectures [27, 33]. For example, [27] proposes an explanation method, particularly for LSTMs. The model-agnostic methods [30, 35, 42] could be deployed across different neural network architectures.

Local versus global: Local methods [35, 42] provide an explanation with respect to a particular input instance, while global methods [13] explain the inner workings of the entire model.

Forms of explanation: The methods also differ in the form of explanations they provide. **Feature-based** methods (e.g., [30, 42]) assign weights to the features depending on their importance to the inference result. **Natural language explanation** methods [11, 28] generate sentences in natural language explaining the reasons behind an inference result. Explanations can also be **example-based**, i.e., training examples that have had the most influence on the inference result for a target input [18].

Interpretable embedding: Since, word embeddings are the building blocks of any natural language processing task, another line of work aims to augment interpretability to the embeddings. These embeddings can then be used as interpretable features that could be helpful in explaining the decision of the model [7, 26, 34].

Game theoretic aspect. [24] utilizes coalition game theory to calculate feature attribution scores. This approach views the features as individual players in a game of prediction, with the payout being the result. The importance is determined by how the payout is distributed among the players (features). The idea has been further investigated in [8, 9, 23, 41].

Quantifying feature interactions. One key aspect that has led to the success of deep neural networks is their ability to learn higher level concepts from lower level features. The importance of feature interactions cannot be adequately captured by the methods mentioned above. The primary goal of the proposed framework is to assign importance scores to these feature interactions or feature groups. Previous attempts at obtaining the importance of feature groups include [3, 5, 14, 22, 27, 37, 43].

3 Method

3.1 Feature Group Attribution Problem

Given a trained DNN model, an input, a baseline, and feature groups that follow a meaningful structure, the goal is to assign each feature group an importance score. Following existing studies, our formulation assumes a baseline b that represents the "zero" input or alternatively absence of a particular feature. In general, feature groups are assumed to have a hierarchical containment structure, which specifies that they can be represented as a directed acyclic graph with the tree being a special case.

3.2 Solution Axioms

The value/importance function (v) that we want to design for assigning importance scores to feature group/set (S) should abide by a set of axioms that we summarize next. Note that we limit ourselves to presenting only the motivation behind the axioms in this article. We refer the reader to [36], for a more formal overview. The first set of axioms are borrowed from cooperative game-theory literature and are termed as *Non-Negativity*, *Normality*, *Monotonicity* and *Superadditivity*. While *Non-Negativity* requires that every feature group has a non-negative value ($v(S) \geq 0$), *Normality* ensures that empty feature group has a value of 0, i.e., $v(\phi) = 0$. *Monotonicity* requires that the value of a feature set is greater than or equal to any of its subsets, i.e., if $S \subseteq T$, then $v(S) \leq v(T)$. Finally, *Superadditivity* requires that the union of two disjoint feature sets is greater than or equal to the sum of the values of the two sets; if $S \cap T = \emptyset$ then $v(S \cup T) \geq v(S) + v(T)$.

Non-Negativity axiom guarantees that each feature has a value/importance score that is not negative, which is intuitively sensible since the value function reflects the importance of a set of features and is inherently directionless. The Normality axiom guarantees that the significance rating given to the empty feature set is zero. In cooperative games, players work together to create the greatest possible value. It is often assumed that a player can always opt to do nothing, resulting in a value of zero. Therefore, if taking action would lead to a negative value, a rational player would always choose to do nothing which forms the basis of the Non-negativity axiom.

We now present three axioms that are reflective of the feature group attribution problem.

Axiom 1 (*Sensitivity*) The axiom consists of two parts. Firstly, a value function must ensure that any feature that has an influence on the model output should be assigned a positive importance score and it naturally extends to any group that this feature is part of. Similarly, any feature that has no effect on the model out should be assigned an importance score of 0 and should not have any contribution to any feature group that it is part of. □

Axiom 2 (*Symmetry Preservation*) Two features are considered *functionally* equivalent if exchanging their values does not lead to any change in the model output. Similarly, two features are *structurally* equivalent if they occupy equivalent positions in the structure established by the feature groups. Symmetry preservation requires that the two features that are structurally and functionally equivalent should have the same contribution to every feature group that they are part of. □

Axiom 3 (*Implementation invariance*) Irrespective of the implementation of a given DNN model, the corresponding value or importance score for each feature group should remain the same, i.e., the scores should not be influenced by the complexity of the model. □

3.3 Our Method: Integrated Directional Gradients

We now present our method Integrated Directional Gradients or IDG. This method is built upon the Integrated Gradients method [42], and is inspired by Harsanyi dividends [10] in cooperative game theory. The key idea is to construct the value function in terms of the "dividends" that come from each feature group. We assume that every feature group contributes an "additional value" to the DNN model, which we term as "dividend" of the group.

For a single feature, the dividend is also its value. However, for a feature group, the value and the dividend are distinct. For our framework, we take the directional derivative of the DNN function in the direction of the given set of features as the measure of the importance of the interaction of a given feature group. Intuitively, a directional gradient quantifies the sensitivity of a DNN function to changes in the input in the direction of a subset of features. To deal with issues such as gradient saturation, we use the absolute value of IDG, represented by the path integral of the directional gradient over the straight line path from the baseline b to the input x to be the dividend of the feature group. Further, the sign of IDG is indicative of the nature of the contribution.

$$z_i^S = \begin{cases} x_i - b_i & \text{if } a_i \in S \\ 0 & \text{otherwise} \end{cases} \tag{1}$$

$$\nabla_S g(x) = \nabla g(x) \cdot \hat{z}^s \quad \text{where } \hat{z}^s = \frac{z^S}{\|z^s\|} \tag{2}$$

$$\mathrm{IDG}(S) = \int_{\alpha=0}^{1} \nabla_S g\,(b + \alpha(x - b))\,\mathrm{d}\alpha \tag{3}$$

Table 1 Terminology table

Symbol	Description
$A = \{a_1, a_2, \ldots, a_n\}$	Set of all features; $\|A\| = n$
$\mathcal{P}(A)$	Power set of the set of all features A
$x \in \mathbb{R}^n$	Feature vector
$b \in \mathbb{R}^n$	Baseline feature vector
$f: \mathbb{R}^n \rightarrow \mathbb{R}$	Deep neural network function
$S, T \subseteq A$	Given subsets of features
$M \subseteq \mathcal{P}(A)$	The set of meaningful feature groups
$x^S \in \mathbb{R}^n$	Feature subset vector for S; $x_i^S = x_i$ if $a_i \in S$ otherwise $x_i^S = 0$
$b^S \in \mathbb{R}^n$	Feature subset baseline for S; $b_i^S = b_i$ if $a_i \in S$ otherwise $b_i^S = 0$
$v(S) : \mathcal{P}(A) \rightarrow [0, 1]$	Value of feature groups
$d(S) : \mathcal{P}(A) \rightarrow [0, 1]$	Dividend of feature groups

$$d(S) = \begin{cases} \frac{|\text{IDG}(S)|}{Z} & \text{if } S \in M \\ 0 & \text{otherwise} \end{cases} \tag{4}$$

$$Z = \sum_{S \in M} |\text{IDG}(S)| \tag{5}$$

$$v(S) = \sum_{T \in \{T | T \subseteq S \wedge S \in M\}} d(T) \tag{6}$$

Equations 1 to 6 defines the method for computing the value/importance $v(S)$ of a feature subset utilizing IDG. The notations used in the above equations are summarized in Table 1.

Importantly, IDG satisfies all the axioms.[1]

The computation of IDG is illustrated through an example sentence in Fig. 1a. The task is binary sentiment classification with the model being XLNet-base. The meaningful features groups are obtained following the sentence's parse tree. We compute the value function in a bottom-up manner following the parse tree structure starting from the leaves (embedding dimension) to the root (sentence).

[1] For detailed proofs refer to the original paper [36].

4 Evaluation

4.1 Setup

For evaluation, we consider sentiment classification task across three different datasets—Stanford Sentiment Treebank (SST) [38], Yelp reviews [46], and IMDB [25]. We train three state-of-the-art models—XLnet-base [44], XLnet-large [44], and BERT-itpt [39].

4.2 "CHECKLIST" Tests

Evaluating any explanation framework poses a significant challenge. It is extremely difficult to conclusively state that any given framework is correct, i.e., the features (words) deemed as the most important ones by the explanation framework are indeed the ones considered by the classifier model to be important. Inspired by the "CHECKLIST" methodology proposed in [31] to understand the behavior of NLP models, we devise our own set of "CHECKLIST" tests as sanity checks for explanation frameworks.

Typically, we design a set of simple template examples like The movie was <quantifier> <adjective>, The actors in the movie were <negation> <adjective>. Note that unlike the reviews present in the datasets, the linguistic structure of these phrases is simple, and no other words except the <adjective> or <negation> (where present) convey any information about the sentiment of the phrase. We argue that given the classifier model makes a correct prediction, it should assign more weight to the <adjective> or <negation> present in the phrase. If the explanation framework makes a similar inference, we can reckon that it is correct. We deploy the model trained on IMDB movie reviews (details discussed later in this section) for inferring sentiment. We illustrate with an example in Fig. 2. We consider two template sentences The movie was brilliant. and The movie was not that good. Note that in the first case "brilliant" is the only word that determines the sentiment of the sentence. Consequently, it is assigned the highest score by our method. Similarly, in the second case, the phrase "not that good" determines the sentiment and is assigned the highest score.

4.3 Correctness

Evaluating the correctness of an explanation, i.e., whether the model indeed considers the features highlighted by the explanation framework to be important, is difficult. A common strategy is to remove the features with high-importance scores and observe the output of the model on this manipulated input. Ideally, the model output should

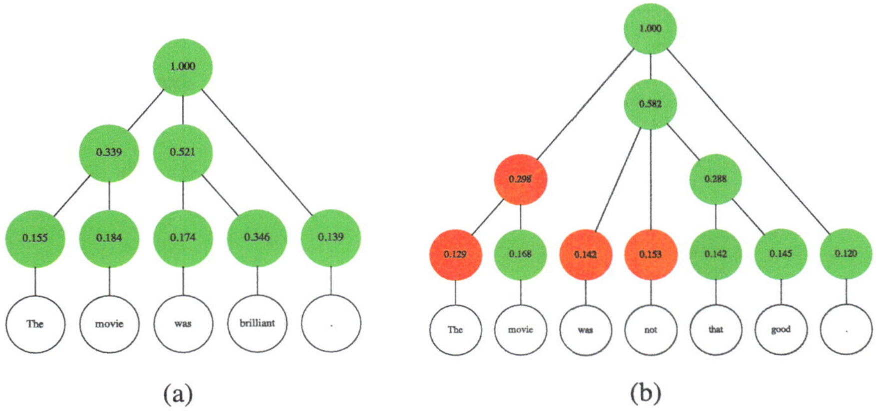

Fig. 2 The scores and the corresponding directions as inferred by our model on a template sentence **a** `The movie was` *`brilliant.`* and **b** `The movie was` *`not that good.`* Red represents negative contribution to the inferred class while green represents positive contribution. We observe that the word "brilliant" has been assigned a much higher score compared to the other words which are not supposed to have influence on the model's inference. Similarly, the phrase "not that good" has been assigned a high value given its contribution to the overall sentiment score. In both cases, we used a BERT-itpt model finetuned on SST dataset

change given the important features have been removed. However, in text, arbitrarily removing words may lead to a sentence devoid of a syntax structure or semantic sense. Since the model had not encountered such an input during training, the output of the model on this particular input may not be trustworthy [12]. We devise a novel way of solving this issue. We consider reviews from Yelp with multiple sentences where the first and the second part are of opposite sentiments. We then utilize a natural language generation model (BART [21]) to generate the second sentence, utilizing the first one as the prompt. We illustrate with an example in Fig. 3. The original review collected from Yelp consists of two sentences with different sentiments. However, the overall sentiment is classified as negative. We use the first sentence as a prompt. The generated second sentence has a positive sentiment. Since the generation model takes cues from the prompt, it is more likely to generate a sentence with a sentiment in line with the first sentence. We hence obtain a pair of sentences with opposite outcome but in line with the semantic and syntactic structure.

We observe that the sentiment in the second case is classified to be positive. Given that the second sentence had a greater influence on the classification result and that replacing it leads to a different outcome, demonstrates the correctness of IDG. We repeated this experiment with other examples obtaining similar results.

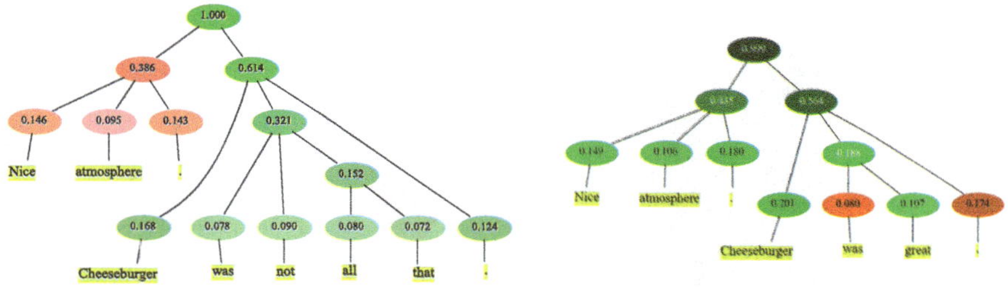

(a) (b)

Fig. 3 **a** The value function scores of the original review `Nice. Atmosphere. Cheeseburger was not all that.` The value function scores of the generated review where the first sentence was used as a prompt. The second sentence is now `Cheeseburger was great.`. In the first case, the overall sentiment is classified to be negative with the first sentence contributing negatively to the overall score. In the second case, the overall sentiment is classified to be positive with both parts contributing positively to the overall score. This demonstrates that the attributions generated by IDG are indeed correct. We used

4.4 Capturing Semantic Interaction

In this set of experiments, we investigate whether IDG is able to capture semantic interactions through negations and conjunctions. For illustration, we consider one example each from SST and IMDB datasets in Fig. 4a and b, respectively. In Fig. 4a, although the first part has a positive sense, when augmented with the second part, the overall sense becomes negative. IDG is able to capture this interaction as demonstrated by the obtained scores. Similarly, the example in Fig. 4b consists of the sentence `I don't understand how the movie receives such a high rating`. While the phrase "high rating" itself has a positive sense, when the entire sentence is considered the sense becomes negative. The obtained importance scores are indicative of this interaction. Additional results and details can also be found in [36].

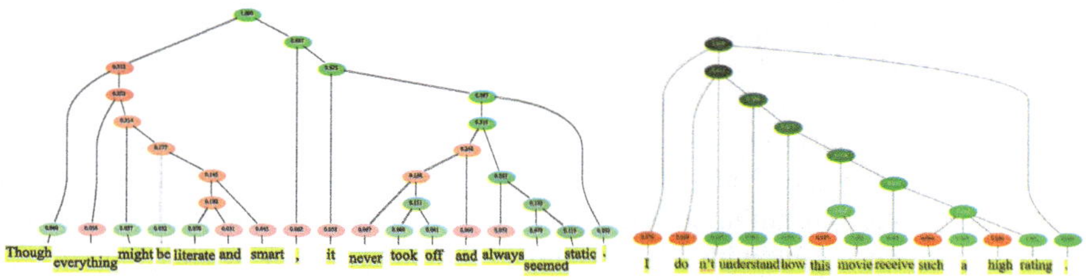

(a) (b)

Fig. 4 Results of IDG when deployed for reviews sampled from **a** SST and **b** Yelp

5 Conclusion

In this work, we investigated the feature group attribution problem. We developed a set of axioms that any solution framework should abide by. We introduced IDG as a solution framework that not only satisfies all the axioms, but is also effective when deployed to real-world datasets and state-of-the-art DNN-based classifier models.

References

1. Bach S, Binder A, Montavon G, Klauschen F, Müller K-R, Samek W (2015) On pixel-wise explanations for non-linear classifier decisions by layer-wise relevance propagation. PLoS ONE 10(7):e0130140
2. Camburu O-M (2020) Explaining deep neural networks. arXiv:2010.01496
3. Chen H, Zheng G, Ji Y (2020) Generating hierarchical explanations on text classification via feature interaction detection. In: Annual meeting of the association for computational linguistics, pp 5578–5593
4. Chen J, Jordan M (2020) Ls-tree: model interpretation when the data are linguistic. In: AAAI conference on artificial intelligence, vol 34, pp 3454–3461
5. Cui T, Marttinen P, Kaski S et al (2020) Learning global pairwise interactions with bayesian neural networks. In: European conference on artificial intelligence. IOS Press
6. Devlin J, Chang M-W, Lee K, Toutanova K (2018) Bert: pre-training of deep bidirectional transformers for language understanding. arXiv:1810.04805
7. Engler J, Sikdar S, Lutz M, Strohmaier M (2022) SensePOLAR: word sense aware interpretability for pre-trained contextual word embeddings. In: Findings of the association for computational linguistics: EMNLP, pp 4607–4619
8. Frye C, Rowat C, Feige I (2020) Asymmetric Shapley values: incorporating causal knowledge into model-agnostic explainability. In: Advances in neural information processing systems, vol 33
9. Ghorbani A, Zou J (2020) Neuron Shapley: discovering the responsible neurons. arXiv:2002.09815
10. Harsanyi JC (1963) A simplified bargaining model for the n-person cooperative game. Int Econ Rev 4(2):194–220
11. Hendricks LA, Akata Z, Rohrbach M, Donahue J, Schiele B, Darrell T (2016) Generating visual explanations. In: European conference on computer vision. Springer, Berlin, pp 3–19
12. Hooker S, Erhan D, Kindermans P-J, Kim B (2019) A benchmark for interpretability methods in deep neural networks. In: Advances in neural information processing systems
13. Ibrahim M, Louie M, Modarres C, Paisley J (2019) Global explanations of neural networks: Mapping the landscape of predictions. In: Proceedings of the 2019 AAAI/ACM conference on AI, ethics, and society, pp 279–287
14. Janizek JD, Sturmfels P, Lee S-I (2020) Explaining explanations: axiomatic feature interactions for deep networks. arXiv:2002.04138
15. Jin X, Wei Z, Du J, Xue X, Ren X (2019) Towards hierarchical importance attribution: explaining compositional semantics for neural sequence models. In: International conference on learning representations
16. Kim B, Wattenberg M, Gilmer J, Cai C, Wexler J, Viegas F et al (2018) Interpretability beyond feature attribution: quantitative testing with concept activation vectors (TCAV). In: International conference on machine learning. PMLR, pp 2668–2677
17. Kim J, Rohrbach A, Darrell T, Canny J, Akata Z (2018) Textual explanations for self-driving vehicles. In: Proceedings of the European conference on computer vision (ECCV), pp 563–578

18. Koh PW, Liang P (2017) Understanding black-box predictions via influence functions. In: International conference on machine learning. PMLR, pp 1885–1894
19. LeCun Y, Bengio Y, Hinton G (2015) Deep learning. Nature 521(7553):436–444
20. Lei T, Barzilay R, Jaakkola T (2016) Rationalizing neural predictions. In: Proceedings of the 2016 conference on empirical methods in natural language processing, pp 107–117
21. Lewis M, Liu Y, Goyal N, Ghazvininejad M, Mohamed A, Levy O, Stoyanov V, Zettlemoyer L (2019) Bart: denoising sequence-to-sequence pre-training for natural language generation, translation, and comprehension. arXiv:1910.13461
22. Liu Z, Song Q, Zhou K, Wang T-H, Shan Y, Hu X (2020) Detecting interactions from neural networks via topological analysis. In: Advances in neural information processing systems, vol 33
23. Lundberg SM, Erion G, Chen H, DeGrave A, Prutkin JM, Nair B, Katz R, Himmelfarb J, Bansal N, Lee S-I (2020) From local explanations to global understanding with explainable AI for trees. Nat Mach Intell 2(1):2522–5839
24. Lundberg SM, Lee S-I (2017) A unified approach to interpreting model predictions. In: Advances in neural information processing systems, pp 4765–4774
25. Maas A, Daly RE, Pham PT, Huang D, Ng AY, Potts C (2011) Learning word vectors for sentiment analysis. In: Annual meeting of the association for computational linguistics: human language technologies, pp 142–150
26. Mathew B, Sikdar S, Lemmerich F, Strohmaier M (2020) The polar framework: polar opposites enable interpretability of pre-trained word embeddings. In: Proceedings of the web conference, pp 1548–1558
27. Murdoch WJ, Liu PJ, Yu B (2018) Beyond word importance: contextual decomposition to extract interactions from LSTMS. In: International conference on learning representations
28. Park DH, Hendricks LA, Akata Z, Rohrbach A, Schiele B, Darrell T, Rohrbach M (2018) Multimodal explanations: justifying decisions and pointing to the evidence. In: Proceedings of the IEEE conference on computer vision and pattern recognition, pp 8779–8788
29. Rajani NF, McCann B, Xiong C, Socher R (2019) Explain yourself! leveraging language models for commonsense reasoning. arXiv:1906.02361
30. Ribeiro MT, Singh S, Guestrin C (2016) why should i trust you? explaining the predictions of any classifier. In: Proceedings of the 22nd ACM SIGKDD international conference on knowledge discovery and data mining, pp 1135–1144
31. Ribeiro MT, Wu T, Guestrin C, Singh S (2020) Beyond accuracy: behavioral testing of nlp models with checklist. arXiv:2005.04118
32. Rudin C (2019) Stop explaining black box machine learning models for high stakes decisions and use interpretable models instead. Nat Mach Intell 1(5):206–215
33. Selvaraju RR, Cogswell M, Das A, Vedantam R, Parikh D, Batra D (2017) Grad-cam: visual explanations from deep networks via gradient-based localization. In: IEEE international conference on computer vision, pp 618–626
34. Şenel LK, Şahinuç F, Yücesoy V, Schütze H, Çukur T, Koç A (2022) Learning interpretable word embeddings via bidirectional alignment of dimensions with semantic concepts. Inf Process Manag 59(3):102925
35. Shrikumar A, Greenside P, Kundaje A (2017) Learning important features through propagating activation differences. In: International conference on machine learning. PMLR, pp 3145–3153
36. Sikdar S, Bhattacharya P, Heese K (2021) Integrated directional gradients: feature interaction attribution for neural NLP models. In: Proceedings of the 59th annual meeting of the association for computational linguistics and the 11th international joint conference on natural language processing, vol 1: Long Papers, pp 865–878
37. Singh C, Murdoch WJ, Yu B (2018) Hierarchical interpretations for neural network predictions. In: International conference on learning representations
38. Socher R, Perelygin A, Wu J, Chuang J, Manning CD, Ng AY, Potts C (2013) Recursive deep models for semantic compositionality over a sentiment treebank. In: Conference on empirical methods in natural language processing, pp 1631–1642

39. Sun C, Qiu X, Xu Y, Huang X (2019) How to fine-tune bert for text classification? In: China National conference on Chinese computational linguistics. Springer, Berlin, pp 194–206
40. Sundararajan M, Dhamdhere K, Agarwal A (2020) The Shapley Taylor interaction index. In: International conference on machine learning. PMLR, pp 9259–9268
41. Sundararajan M, Najmi A (2020) The many Shapley values for model explanation. In: International conference on machine learning. PMLR, pp 9269–9278
42. Sundararajan M, Taly A, Yan Q (2017) Axiomatic attribution for deep networks. In: International conference on machine learning. PMLR, pp 3319–3328
43. Tsang M, Rambhatla S, Liu Y (2020) How does this interaction affect me? interpretable attribution for feature interactions. In: Advances in neural information processing systems, vol 33
44. Yang Z, Dai Z, Yang Y, Carbonell J, Salakhutdinov RR, Le QV (2019) Xlnet: generalized autoregressive pretraining for language understanding. In: Advances in neural information processing systems, pp 5753–5763
45. Yoon J, Jordon J, van der Schaar M (2018) INVASE: instance-wise variable selection using neural networks. In: International conference on learning representations
46. Zhang X, Zhao J, LeCun Y (2015) Character-level convolutional networks for text classification. In: Advances in neural information processing systems, pp 649–657

SPRINGER NATURE

GPSR Compliance

The European Union's (EU) General Product Safety Regulation (GPSR) is a set of rules that requires consumer products to be safe and our obligations to ensure this.

If you have any concerns about our products, you can contact us on ProductSafety@springernature.com

In case Publisher is established outside the EU, the EU authorized representative is:

Springer Nature Customer Service Center GmbH
Europaplatz 3
69115 Heidelberg, Germany

The manufacturer's authorised representative in the EU is Springer Nature Customer Service Centre GmbH, Europaplatz 3, 69115 Heidelberg, Germany. If you have any concerns regarding our products, please contact ProductSafety@springernature.com

Printed and bound by CPI Group (UK) Ltd, Croydon, CR0 4YY
01/10/2026
02225875-0007